Where does the space end?

Is the universe infinite?

Werner Kinnebrock

Copyright © 2018 Werner Kinnebrock

All rights reserved.

CreateSpace Independent Publishing Platform

ISBN: 1719567581
ISBN-13: 9781719567589

CONTENT

	Introduction	5
1	Newtons space, the space of life	8
2	The empty space	16
3	The geometry of space	22
4	The geometry of nature	36
5	Einsteins space-time	45
6	Curved spaces	56
7	The universe	62
8	Ist the space infinite?	83
9	Holes in space?	89
10	The anthropic universe	95
11	Spookiness in space	102
12	Mathematical spaces	107
13	A brief history of space	115
14	Space-speculations	120
	Appendix	126
	Literature	131

Introduction

Vacuum energy permeates the entire room. Attempts have been made to explain the dark energy of space which affects the acceleration of the expansion of the universe by the vacuum energy inside the quantum theory. (see. Chapter 7). However, the calculations showed that the energy of the vacuum is much too large to explain the dark energy.

Space is the stage on which is going on our lives. Along with the time it forms the framework within which we operate. As time progresses inevitably, the space offers us freedom, because we can decide our way free. Even in painting, in sculpture and in architecture we use the freedoms in the room.

At the same time space and time are a limit to us, because we can not leave the framework of time and space.. Even our thinking is only possible under the provisions of space and time, as the Konigsberg philosopher Immanuel Kant already recognized 1751.

We find unimaginably huge rooms in the vastness of the universe. There are voids (empty space) with a diameter of over one hundred million light years. In the constellation Eridanus an empty space (void) was discovered with a diameter of about one billion light-years. But not only there: Matter consists almost entirely of empty space. Atomic nuclei and electrons, the building blocks of visible matter, account for less than 0.001 percent of the atom. If we remove all matter from the atoms of a skyscraper, a lump in the size of a rice grain would remain. The rest would be empty space. This lump could have a weight of several hundred million kilograms.

What is space? Why are there precisely three spatial dimensions and not four or six? Is space real or just nothingness? These questions are so difficult to answer as the question of the time. Space is one of the greatest mysteries of physics. Suppose that we would remove from a room all matter until nothing is left. Is there still room? Of course, because it is defined by its limitations, such as the walls of a container. Now we remove these. If no limitations are there, the space is infinitely extended. But since it contains nothing we can no longer perceive it. Is room still available or is it gone? What comes first, the space or matter in it?

These questions were already occupied by the ancient Greeks. At that time arose as a spatial calculation method Euclidean geometry. It goes back to Euclid and was a model for scientific formal systems. Based on axioms there arose developed statements that are derived through logical proofs. For centuries, almost all the sciences have oriented to the logical structure of Euclidean geometry.

From Antiquity to the Middle Ages space was conceived as a kind of container in which you saved something. One speaks of an absolute conception of space. Isaac Newton relied on this absolute space and an absolute time notion when he developed his system of mechanics, which revolutionized physics.

Later, you looked at space as a space between the material bodies (relational understanding of space). Representatives of this conception of space were Gottfried Wilhelm Leibnitz and the physicist Ernst Mach.

Albert Einstein revolutionized the view of space by the concept of relativity. From now on you looked at a four-dimensional variety, consisting of the three dimensions of space and time coordinate, called spacetime. The room was now no longer static and absolute, but could be variable depending on movements.

Ten years after this discovery Albert Einstein expanded his

concept of space, by presenting a theory in which the space is curved like a spherical surface is curved. This curvature is caused by matter and energy in space.

The quantum theory influenced the conception of space by developing ideas, which Albert Einstein did not believe. He called them "ghostly". But the validity of these ideas was proved later. Thus, two elementary particles that are billions of kilometers away from each other, "know from each other" in the sense that if one changes the measured value, the other at the same moment responds.

We try to reflect the structures of the space in the statements of geometry. Here the objects of geometry are such as circles, ellipses and idealized objects, which are not found in nature in this pure form. The Geometry of Nature arises in shapes as we find it in snowflakes, leaves and mountain silhouettes. It is a fractal geometry, characterized by self-similarity of objects. Computer procedures can produce fractal structures, which Benoit Mandelbrot discovered a few decades ago. This produces images of unimaginable beauty and harmony.

No science is concerned so much with the concept of space as mathematics. Here we have three-dimensional spaces, four dimensional spaces or eight or generally n-dimensional spaces of arbitrary dimension n. Sets of functions or sets of numbers can have spatial structure and are treated as spaces. Curved spaces like the sphere are objects of investigation. One application would be the curved space of Einstein.

The room might have many surprises that are yet to be discovered. We still do not know definitively what space really is.

This book attempts to represent all the described properties of space generally understandable.

1. NEWTONS SPACE, THE STAGE OF LIFE

I do not like spaces having been designed by indoor architects. I always feel like I'm sitting on the stage- the curtain opens and I can not get my text..

Sir Peter Ustinov, actor, 1921-2004

What is space? The physicists can describe properties and effects of the space, but what space and time are ultimately, there is no satisfactory answer. Why is the space in three dimensions, not four or five-dimensional? Consists the room of subparticles how the atoms? Is space real or just nothingness? Why is there at all "space"? Questions about questions to which there are no answers.

1.1 Space as we experience it

Space is the stage on which is going on our lives. If we add the time, we have the framework that determines our lives and we can not leave our space and time. We are cooped up in a grid that shapes us, gives us a wide range of possibilities, but also specifies limits. Our knowledge is bounded to spatio-temporal images and limited so as Immanuel Kant already recognized 1781

Our overall plans are subjected to the coordinates of space and time. Without time there would be no music and without space no painting and no sculpture. Time and space are the setting for music and dance. Objects we can arrange one above the other or one behind the other, because the space has three dimensions. When we travel, we traverse the space and find out the possibilities offered by the world to us. We determine our path we want to go and thereby acquire an incredible freedom. This freedom is not the freedom of the animals, followed by the only instinct, but is based on decisions from which we form our personality. Space and time are therefore the conditions for the possibility of development of the individual personality structure and thus also the culture.

We linger in memories, or we plan for the future, we may leave space and time in our thoughts. The here and now is replaced by there and yesterday/tomorrow and by there or there. The imagination knows no bounds when we fly in the imagination through space and time, drive to distant lands or when we set forward the time or reset.

Space is everywhere and supports our world. There arise many questions: Why are there room and space? Why is space three-dimensional and not four or five-dimensional? What is space? Is the room made in subparticles like molecules? Is space real or is space just nothingness?

1.2 Properties of the space

What is space? Physicists have no clear answer. Our life is determined by space and time. Let us give solid space coordinates such as latitude and longitude, as well as an accurate time, then a point is precisely determined, in this case, a point on the earth at a particular time. This point is clearly, there can not be a second point with the same coordinates anywhere else.

Thus we have a first property of space and time: you can both fix precisely by coordinates. The space has three dimensions and thus three coordinates, for the time we need only one coordinate, namely time itself.

If we knew all the coordinates of all sorts of matter in the universe and their velocities and accelerations, could we then - at least theoretically - determine the future behaviour of all matter particles and thus calculate the future? This was at least the opinion of Pierre Simon de Laplace, when he wrote in 1776: *"If there is someone who would be able to know for a given moment all the forces of which nature is moved, ..., nothing would be more uncertain for him and the future and the past would be present before his eyes "*

Today we know that this statement is no longer true. Quantum theory and chaos theory showed that the detection of all the tangible realities is out of focus and the impact of movements must not be predictable.

An important characteristic of the room is its locality. If I throw a stone into a lake, then he is making waves near the point of impact. These waves are spreading. The stone affects the lake just nearby, so locally. Every action has consequences within the space in the neighbourhood and propagates outwards. The propagation velocity is always less than the speed of light. The location of the room was therefore a natural property that resulted from the prevailing laws of physics.

There was a lot of excitement when some quantum physicists claimed that in micro-space the location does not apply. Nonlocality should be given allegedly in operations in the microcosm. That would be as if someone threw a stone at Oxford in the River Thames and its waves arise at the time of impact not in Oxford but in the Thames near London. Such an operation would be like Spooky, and Albert Einstein called this process actually as "spooky". He refused to recognize non-locality in the microcosm and remained until his death.

Only in the eighties of the last century, experiments showed that the non-locality in quantum physics actually exists. We will return to this in Chapter 11.

1.3 Space in the atoms and in the universe

In the context of space, we tend to think about the space around us. In the universe there is space in abundance, there are unimaginably large void spaces in eternal darkness. Would we stay in these empty spaces, we could see no stars and no light. Stars and galaxies act there as minor contamination.

But not only in space, we see space in volumes, the atoms consist almost entirely of empty space. If one were to remove from the atoms of the water in a swimming pool with normal size all empty space, so that only the material part of the atom remains, there is only a mass of less than the size of a grain of rice left. This "grain of rice", however, would weigh more than a thousand tons. So we swim in the (almost) empty space, if we take our bath in the swimming pool. Only the uniformly distributed mass of the "rice grain" holds us.

The surrounding space is filled with air, the sea with water and the space of the universe with energy. The entire cosmos is probably filled with a medium that scientists call Higgs field and that gives elementary particles their mass. If we remove all these things there remains the vacuum. But even the vacuum is filled with seething energy of particles that constantly arise and annihilate each other again.

1.4 Newton's Space-Time

Although we can not see, smell or hear the room, it was real for Isaac Newton. The room was the stage of world history, a stage that worked as a stage for what is happening.

Philosophers have debated the space for centuries. So Rene 'Descartes had claimed that space is always relative to the movement. So a passenger on a ship is in a dormant space when sitting in the main cabin. Nevertheless, the room is relative to the coast, because the ship moves. A place in the world is only relatively dormant, because the earth moves. An absolute resting space which includes everything does not exist. This was the opinion of Descartes.

Not so Isaac Newton. He distinguished between the absolute and the relative space. In a script appearing in 1687 he wrote: "An absolute space, of no relation to anything exists. It always remains the same and is motionless.

Isaac Newton's room was pioneering. Newton was born in 1643 in a village named Woolsthorpe. His father died young, and his mother moved to the neighbouring village, where she left the boy with his grandparents, who ran a small farm. It turned out early that the boy did not seem suitable for agriculture. He was able to visit schools through mediation of a relative and in 1660 he was a student at Trinity College in Cambridge. He attended lectures in the natural sciences and in theology and learned Hebrew, Greek and Latin. Later, Newton was Professor in Cambridge. His teacher Isaac Barrow, one of the best mathematicians of that time, had dispensed in favour of Newton on his position because he thought Newton more capable. Newton remained about 30 years in Cambridge before moving to London where he became Warden of the Royal Mint and served as president of the Royal Society. He died there in the year 1727.

In his work "Philosophiae Naturalis Principia Mathematica", which appeared in 1687, Newton laid the foundations for a new approach to the description of mechanical systems as well as the heavenly bodies. The German translation of this basic work was published by J. Wolfers 1872 under the title "Mathematical Principles of Natural Sciences".

Newton saw the room in relation to the matter. The room itself was absolutely, as well as the time. It was as if in the cosmos an oversized clock dictates the time for all points and stars of the universe. The stars were embedded in an immobile immutable space which makes up the universe. His theory of space, time and matter was for centuries an indication and is still regarded as a valuable approximation to what we today call "space". Only by Albert Einstein the concepts of absolute space and absolute time were questioned.

An important mathematical discipline to describe the relationship between space and matter was the differential and integral calculus. This made it possible to describe the relationship between space and matter satisfactorily.

Independently this important mathematical theory was developed by Isaac Newton in England and Gottfried Wilhelm Leibnitz in Germany and both of them used different formalisms. The so-called priority dispute led to clashes between the German, French and Swiss supporters of Leibnitz and Newton's English supporters. Each side accused the other of plagiarism. Only later could be demonstrated that both had developed independently of each other the important discipline of differential and integral calculus.

Using this mathematics now it was possible to describe orbits of celestial bodies and trajectories in space very accurately. Newton first described the gravitation when he wrote::

"It seems ....that there is a force that bodies by virtue of their mere presence through space exert on each other. .... This wonderful effect of things may be called Gravitation"

Newton found that this gravitational force is proportional to the product of the two masses of the attracting bodies divided by the square of the distance. (law of gravitation). He writes: "Gravitation must be caused by an agent acting constantly according to certain laws,". This action he suspects through the work of an omnipresent Creator.

Before Newton, Johannes Kepler had discovered in Prague three important laws affecting the planetary orbits. It turned out that these laws can be derived directly from the law of gravitation of Newton, which proved to be a brilliant confirmation of Newton's law of gravitation.

1.5 Newton and the consequences

The absolute space and absolute time as postulated by Newton led the people in a changed view of the world. Up to now, religiously embedded ideas were starting points of all reflections, but there was now something that seemed to have an absolute existence outside this imaginary world and this was the starting point of absolute thoughts and reflections. Space, time and matter seemed to be basic variables.

In the period after Newton, the science developed on the basis of space, time and matter, where space and time were the absolute and fixed reference values. The calculus discovered by Newton and Leibnitz allowed calculation of the interaction between space and matter. Based on these calculations accurate predictions could be made, corresponding to the reality. The Newtonian world view was common property and bribed by its uniform conception.

In the following years after Newton the success of the new method of calculation resulted to euphoria. The feasibility of all things and the autonomy of the people were the future vision and the reason has been declared as the highest principle. One of the admirers of Newton was Voltaire, who travelled specially to London to meet Newton. When he reached London, he learned of the death of Newton and even took part in the funeral. For him, Newton's life's work was the way finally to find the true knowledge. Newton did - as Voltaire – not connect the science to the theology and made it so mature. However - what Voltaire overlooked - Newton wanted to collect only knowledge and not embark on a new philosophical direction.

According to Goethe the natural sciences will secrete from humanity by Newton's ideas and this he regarded as a great calamity. He feared that the natural sciences could become independent in the art. Hegel later subscribed to this view. The quantum physicist Werner Heisenberg put it this way: "It was not so much interested in nature as it is, but it turned rather the question of what you can make of it. Natural science was transformed into technique ".

The overvaluation of Newtonian statements you can feel when you hear Laplace saying that future generations of astronomers only remains to pick up "remnants of a feast" such as the cataloging of new stars, discovering new comets, etc. Everything else and the essence were already discovered. He adds. "Just as there is only one universe that requires explanation, so no one can do for a second time, what Newton did, the happiest of mortals.

Laplace could not have imagined that 120 years later, Albert Einstein replaced the Newtonian space through the curved space. Space and time, as Newton saw them lost their absoluteness.

2. THE EMPTY SPACE

*Space is void and nothingness.
And because there is nothing
from the mind is covered, in this
space is tremendous energy*

Jiddu Krishnamurti, Indian philosopher

What is vacuum? Vacuum is defined as a completely empty space, a space without matter, without energy, without radiation. However, the quantum physics showed that an empty space is full of energy. Physicists speak of vacuum energy. Later, the existence of vacuum energy has been demonstrated experimentally.

In the following years it developed the idea that permanent virtual particles are produced in a vacuum, arising briefly, but immediately (with their antiparticles) destroy again. So there is a vacuum eternally bubbling like water in a pot of boiling water. One speaks of the vacuum fluctuation or the zero-point energy.

2.1 History

Leucippus and his pupil Democritus were probably the first who coined the term of the vacuum. For them, there was the matter of indivisible particles (atomoi) that moved in empty space. Only the emptiness of space allowed them to move.

However, for Aristotle movement was inconceivable without

a movement cause. Therefore, he and his students participated in a medium that fills the entire space and was called ether. There was the idea that creation is incompatible with the idea of an absolutely empty space, and that nature "abhors" a vacuum (horror vacui). Aristotle was in the Middle Ages as an authority, so that his idea about the vacuum was to the Middle Ages.

At that time it was believed that the vacuum sucks the matter, so it fills up fast, for example, with air. Later it was realized that it is exactly the opposite: matter is pressing into the empty space.

Rene Descartes started from the following consideration: Matter has extension and vice versa there is also no expansion without matter. He concluded that the vacuum must be filled with matter. 400 years later it was discovered that the vacuum is filled with matter and energy.
(see. chap. 2.3).

After Descartes a dispute arose over the question of whether a vacuum is even conceivable. Many contemporaries thought that vacuum is produced by reducing the air pressure. Evangelista Torricelli believed to have created an empty space in this way. Blaise Pascal said however: "Rather tolerates nature its downfall than the smallest empty space".

Until the 19th century it was believed the outgoing thesis of Aristotle that the vacuum is filled with a substance, the ether. In the late 19th century it was assumed that the light in space does not move through an empty vacuum but through the medium of the ether, which fills the entire universe.

The physics professor Albert Abraham Michelson and chemist E.W. Morley tried in vain to prove the existence of an ether. In 1905, a physicist from the Federal Patent Office Bern with the name of Albert Einstein went out from the simple assumption that there might be no ether and the light moves through a pure vacuum. This was the birth of the theory of relativ

2.2 The Magdeburg hemispheres

We take a metal ball and evacuate using a vacuum air pump all the air out of the ball. The ball then there is (almost) a vacuum, so empty space. What laws govern the empty space? Give these laws information on the structure of empty space?

1657 Otto Gericke wanted to know how a ball containing vacuum behaves. Otto Gericke was one of the four mayors of Magdeburg in Germany. He invented a spectacular experiment for which he was knighted by Emperor Leopold I. He called himself then Otto von Guericke, where he put a u in his last name, so that he could be properly addressed by foreign guests, because French was the dominant language in Germany of these days.

Guericke put in summer 1657 two hemispheres made of copper provided with a seal together into a ball and pumped out the air from the inside. Then before each hemisphere eight horses were harnessed, which should try to pull apart the hemispheres again. He did not succeed. As the balls were then refilled with air, they fell apart. The pressure of the surrounding air to the vacuum ball prevented a breakup.

Guericke's contemporaries were impressed by this experiment, the term "Magdeburg hemispheres" became known.

2.2 What is a vacuum?
At the hemispheres of Magdeburg it was the external air pressure that held the two halves. A vacuum is usually understood as a spatial portion whose inside air pressure is substantially smaller than the standard air pressure.

We consider below a space that is completely empty, the pressure of which is null. If we want to establish such a completely empty space, we must first remove all atoms and molecules. Then it yet contains electromagnetic waves such as light and heat radiation. So we will shield any radiation. Then there remains the thermal radiation of the container walls. In order to prevent this, we have to cool down the tank to the absolute zero, ie somewhat below -273 degrees C. This is practically impossible, but then we come pretty close up to this form of the vacuum.

Could we cool to absolute zero, we would have a true vacuum. Could this space give information to us what space is ultimately?

2.3 The vacuum is not empty
Max Planck held in 1900 a lecture which revolutionized physics and founded the quantum physics. When light is incident on a body, it reflects part of the light and absorbs the rest. Bodies which do not reflect and absorb the radiation are black bodies. The absorbed light energy is transformed into heat and is emitted as thermal radiation.

At the end of the 19th century, physicists have found formulas for the energy distribution of the radiation of a black body, but these formulas were inaccurate. Therefore

Max Planck suggested in 1900 a formula that represented the distribution right. Since the intensity of the radiation depends on the temperature of the black body, the temperature was an integral part of Planck's formula. You just have to use the temperature T in the formula and obtain the correct energy distribution.

It was surprising that even for the vacuum temperature $T = 0$ the formula provided energy values. Accordingly, an absolutely empty black body would also radiate energy. If the formula is correct, even energy occurs in a vacuum, the physicists describe this as zero point energy.

Planck did not believe so completely in his formula. He described it as a transition which might be superseded sometime by a better formula.

But this was never the case. The formula of Planck clearly shows that the vacuum would contain energy, so it is never completely empty.

1948 the Dutch physicist Hendrik Casimir suggested an experiment that should demonstrate the vacuum fluctuation: Two metal plates should be placed side by side in parallel in a vacuum. To turn off annoying heat rays, both plates and the surrounding area should be cooled down to the absolute temperature zero. If there is the zero-point energy in this vacuum, particles constantly between the plates emerge and destroy again.

We know that particles are described by waves, that is, every particle is simultaneously a wave. Are the plates very close together (eg a few nanometres) located, some waves between the plates can no longer arise because the plate spacing is too small. But these particles are outside the plates. As a result, there are more virtual particles outside the plates than in the interior. This leads to an external pressure which compresses the plates. This pressure should be measurable and this effect is referred to as the Casimir effect.

1957 and 1958 the Casimir effect could be demonstrated experimentally.

One possible explanation for the vacuum energy is that there occur so-called virtual particles and they destroy again. Virtual particle pairs are particle-antiparticle that exist only briefly and thereafter cancel each other out. Here are antiparticles of antimatter particles. When antimatter meets ordinary matter, both annihilate and turn into energy. Permanent are particles and antiparticles created and they annihilate each other. It is a single seething as in a boiling kettle.

Vacuum energy permeats the entire room. Attempts have been made to explain the dark energy of the space which effects the acceleration of the expansion of the universe, through the vacuum energy inside the quantum theory (see. Chapter 7). However, the calculations showed that the energy of the vacuum is much too large to explain the dark energy.

3. THE GEOMETRY OF THE SPACE

> Training of measurement with the
> ruler and compass, in lines,
> planes and bodies
>
> The first German textbook
> of the perspective geometry written,
> by Albrecht Dürer (1525)

The man at the centre of nature is compelled to dominate nature and to dispose of it. Since he can not overlook the nature in its entirety he partitioned the manageable part and so he made it understandable.

In view of the space he dismantles the space, recognizes and formulates its laws in order to make more calculations. The result is an edifice of thoughts which we call geometry.

The formulated geometry is based on human comprehensibility. In addition, there is a geometry of nature whose beauty we can learn by useing computers. This type of geometry is content of chapter 4. The present section focuses on the geometry that we use daily.

3.1 Beginnings of Geometry

The geometry allows the mastery of space. It researches and formulates the laws that structure the room. At the beginning of the geometry is measuring.

The beginnings of geometric measurement are probably 4000

years ago in Egypt, where after each flood of the Nile the fields had to be re-measured. For science, the geometry developed by Greek scholars such as Thales of Miletus and Euclid.

The famous "Pythagorean theorem", perhaps the most important theorem of geometry, was probably not from Pythagoras, but was known before him. Pythagoras only taught this theorem, therefore it was named after him. A simple and easy proof for this most important theorem you find in chapter 4.2

The mathematician, astronomer and philosopher Thales of Miletus lived from about 624-548 B.C.. and was counted among the seven sages. He was able to predict a solar eclipse exactly.

Thales probably had no teacher and he is at the beginning of western philosophy. The foundations of geometry he learned from the Egyptians and the famous "theorem of Thales' (Fig.1) he took over probably by the Babylonians.

The first, who summarized the mathematical knowledge of his time, was probably the mathematician Euclid, born in Athen. He lived from about 360 BC to 280 BC and was a student at the Academy of Plato. Later he taught in Alexandria. He wrote 13 textbooks, which were known as "elements". They include all aspects of ancient mathematics, including the then-known statements on geometry. Many sets of elements are not derived from Euclid. His power is to have these summarized ordered. He structured his works through definitions, postulates, and axioms sets, where he derived the rates by a rigorous argument. The elements are the most successful mathematics books of all time and have been used

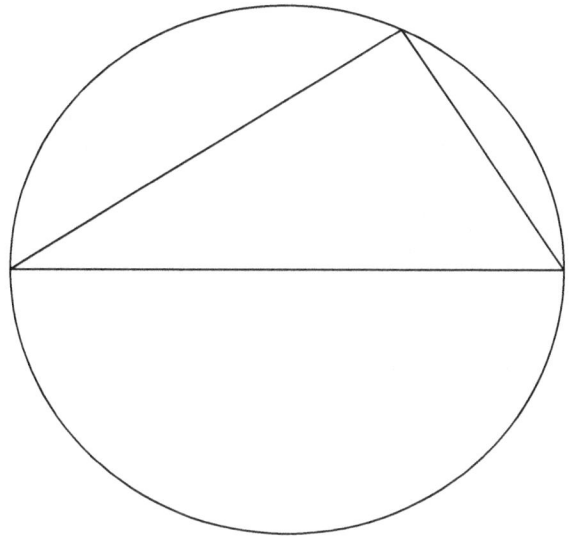

Fig.1: Theorem of Thales: In a triangle in a semi-circle the angle at the top is a right angle.

in England in the 19th century as textbooks. They are based on the geometry that we today learn as school geometry with their sets of triangles, lines and circles, and which is called "Euclidean geometry". It is the intuitive geometry of the plane and the space.

Another important mathematician of antiquity was Archimedes, who lived from about 287 BC to 212 BC in

Syracuse in Sicily. His performance based next to the mathematics of physics, where he first formulated the law of leverage mathematics of physics. His constructed throwing machines and these were used in the defence of Syracuse during the second Punic War against the Roman siege. The major impetus for shipbuilding (Archimedes' principle) he is said to have found in the bath and he ran naked into the street with the exclamation, "I have found it" with joy.

Archimedes was able to prove that the circumference of a circle to its diameter behaves just like the circular area for square to the radius. The calculated value is what we call the number π today. Archimedes was a guide on how you can calculate this number with arbitrarily high accuracy.

"I beg you, don't destroy my circles," Archimedes said to a soldier when he was engrossed in his in sand drawn figures and the Roman soldier in the conquest of Syracuse in 212 BC invaded in his house. Then the soldier killed him. This was reported by the Roman writer Valerius Maximus.

3.2 Euclidean geometry
Euclid was the first who succeeded in a strictly hierarchical building of the geometry according to fixed rules and laws. Every mathematical statement such as the Pythagorean theorem or the theorem of Thales must be proved. For a proof proven records are used. The proof of these propositions in turn leads back to even simpler sentences, and by continuing the process in such a way you find basic geometrical statements that are irreducible and represent the basic building blocks of the theory. These basic building blocks are called axioms. Axioms are statements that arise from the perception and are spontaneously considered "correct". They are not provable. Some axioms of Euclidean

geometry may be mentioned:

1. Through a point outside a straight line you can exactly draw a parallel to the straight line (Parallel Axiom)

2. If there are two different points P and Q, then there is exactly one line containing these points.

3. For each straight line there exists at least one point out of the line.

From the axioms one can draw conclusions, and since prove geometric sets. Some of these theorems are for example:

1. The sum of angles in a triangle is 180 degrees

2. In a right triangle, the sum of the squares of the hypotenuse is equal to the square of the cathetus. (Pythagorean theorem)

3. The angle in a semi-circle is a right angle (see Fig. 1)

All the so derivable sets from the axioms form the Euclidean geometry. It is a geometry of the plane and the space. All geometric calculations in geodesy, architecture, technical development etc. are based on Euclidean geometry.

Until the 19th century the Euclidean geometry was considered as a model for a structurally correct formulated science. But then it became apparent that the formal structure of the geometry, as Euclid defined it, had gaps (see sect. 3.3).

GEOMETRIC PROOF OF THE PYTHAGOREAN THEOREM

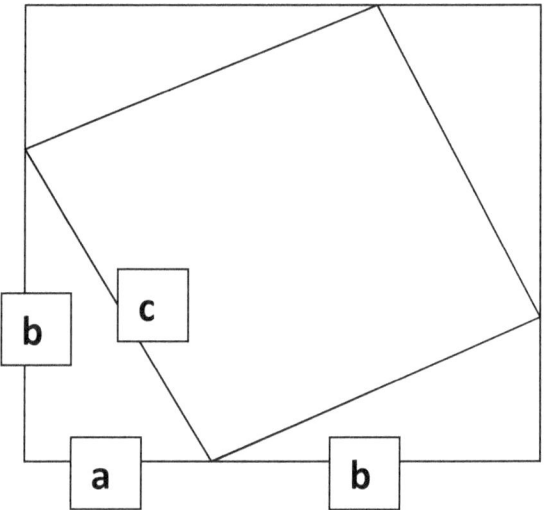

In the outer square is an inner square located.

The surface of the outer square
$$A = (a + b)^2$$

> *On the other hand, the area arises from 4 triangles with the surfaces of each ½ a * b and the surface of the inner square c^2, so*
>
> $$A = 4 * ½ * a * b + c^2$$
>
> *Equating results in*
>
> $$A = (a + b)^2 = 4 * ½ * a * b + c^2$$
>
> *or*
>
> $$A = a^2 + 2ab + b^2 = 2ab + c^2$$
>
> *it follows:*
>
> $$a^2 + b^2 = c^2$$

The German mathematician David Hilbert formulated a correct geometry based on axioms, which he published in 1899 in his book "Foundations of Geometry".

3.3 The discovery of non-Euclidean geometry

It was long believed that Euclidean geometry is the only possible geometry, because it describes all observable phenomena. The described geometry of Euclid and the physics formulated by Newton were the foundations, whose performance had proved in all applications. They were such a thing as absolute truth.

In 1830 a Hungarian officer and a Russian mathematician found a new geometry that differed from the Euclidean geometry. The starting point was the above mentioned first axiom, the parallel axiom:

Axioms - ie the basic statements of a mathematical theory - have the property that they are independent. If they were dependent one could prove one of the axioms using the others and so it would no longer be an axiom, but a derivable sentence. It is therefore necessary, for a mathematical theory, which is based on axioms, to prove the independence of the axioms.

Since Euclid mathematicians suspected that the axioms of Euclidean geometry are not independent. The proof of independence could not be provided. Specifically, it was believed that the parallel axiom is derivable from the other axioms (the first axiom in the upper section). Both in ancient times and in the Middle Ages the mathematicians involved in both the Greek as well as in the Arab world with this problem. A solution they did not find. Even the famous mathematician Carl Friedrich Gauss (1777-1860) was interested in this issue, calling it a "continuing scandal of mathematics".

A solution found in 1830, the Hungarian officer Johann Bolyai (1802-1860) and the Russian mathematician Nikolai Lobachevsky (1792-1856). Their reasoning was the following:

If one would falsify one of the axioms of a theory und so change its statements, then it would arise a new theory or - if the axiom depends an the others - it would lead to contradictions.

An example may illustrate this: We consider three axioms:

Axiom 1: a is an element of the set A (ie a is in the set A)
Axiom 2: a is a member of the set B (ie b is in the set B)
Axiom 3: a is a member of the intersection of A
and B (a is both in A and in B)

One sees immediately that Axiom 3 depends on Axiom 1 and Axiom 2. We change Axiom 3 to the new statement:

a is no element of the intersection of A and B

The immediate results in the contradiction that a either can not be in A or a is not in B and this contradicts axiom 1 or axiom 2. This reveals that Axiom 3 is not independent.

If therefore the parallel axiom is really a dependent axiom, its distortion would lead to a contradiction. The Russian mathematician Nikolai Lobachevsky (1792-1856) and the Hungarian officer Johann Bolyai (1802-1860) studied the problem and falsified the parallel axiom

Through a point outside a line there is exactly one parallel line

by the statement:

Through a point outside a line there is more than one parallel line

The expected opposition they did not find, however, they realized very soon that they had found a completely new geometry with strange properties. So the triangle had not a total angle of 180 degrees, but less than 180 degrees in this new geometry. To a point outside a line there were an infinite number of different parallel lines. This geometry was completely different from the only known Euclidean geometry. Today, this geometry is referred to as a "hyperbolic geometry".

Later, the German mathematician G. Riemann replaced the parallel postulate by the statement:

Through a point outside a line there is no parallel line

and found another geometry. In this geometry, the triangle has an angular amount that is greater than 180 degrees. It is referred to as an elliptical geometry.

The spherical surface turned out to be elliptical. A triangle on the spherical surface has is in fact a sum of angles greater than 180 degrees. Otherwise, the new geometries were pure mind games, as the world around us seemed to be Euclidean.

As Albert Einstein in 1916 his General Theory of Relativity introduced, the newly discovered geometries were suddenly to date. It turned out that the geometry of the universe may well be elliptical and hyperbolic. Whether it is hyperbolic, Euclidean, or elliptical, depends on the density of matter. Today's measurements, however, suggest that the universe may be Euclidean.

3.4 The elliptical World

Suppose our universe has an elliptical geometry. In this case, the universe were curved everywhere as a ball-surface. The spherical surface is not Euclidean, but elliptical. The triangle on the spherical surface has, for example, a sum of angles greater than 180 degrees (see Fig.2).

Where does the space end?

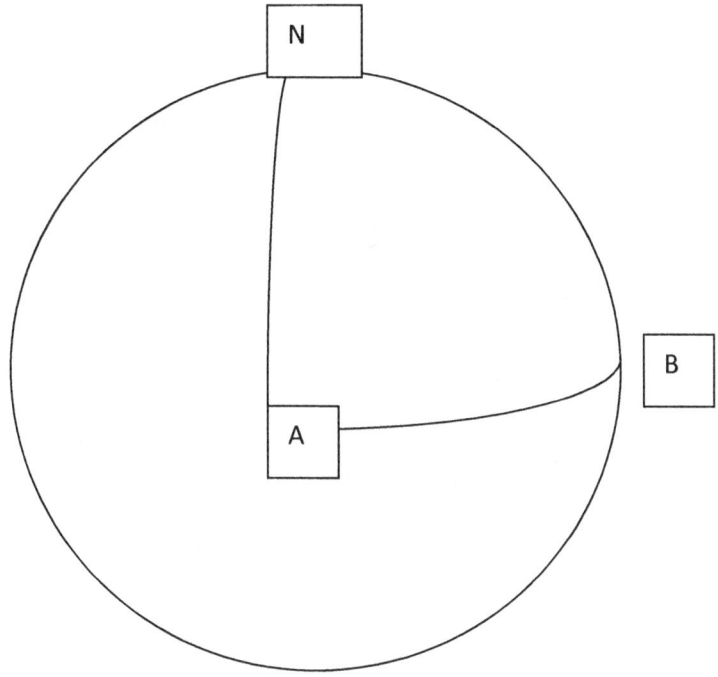

Fig.2: The triangle on an eighth of a sphere has the sum of angles 270 degrees, since the corner angle of 90 degrees are in the points N, A and B

According to the equations of general relativity of Einstein, it is quite possible that the geometry of a space is elliptical. How is this geometry to understand?

The Earth is a three-dimensional sphere, but its surface is only two-dimensional to the two directions of north-south and east-west. In this sense we can say: A ball is three-dimensional, its surface is two-dimensional.

For mathematicians, it is no problem to represent a sphere, which is four-dimensional. Do not try to imagine this four-dimensional sphere. You will not succeed. The reason: Our imagination is space-oriented, in this space we live and this space (room) is unfortunately only in three dimensions. We can only think in three dimensions. Of course, we can also think in two-dimensions (plane) or in one-dimensions (line), but four-dimensional? Impossible.

The "surface" of a four-dimensional sphere has one dimension less, so it is three-dimensional and so it is a room in our sense.. It might be possible that we live in such a room. This space would be elliptical. Just as a spherical surface is curved, such a space would be curved and we would not perceive as inhabitants the curvature of this world.

However, there is a mathematical method to determine if we live in a curved space. We only need to measure a very large triangle and identify its angle sum. If this is greater than 180 degrees, our world would be elliptical and not Euclidean.

The German mathematician Carl Friedrich Gauss was in 1820 commissioned to measure some newly acquired rural area of the Kingdom of Hanover. The measurement was carried out by triangulation, ie the division of the landscape into measurable triangles. As part of this work Gauss measured a particularly large triangle, but he found no deviation from 180 degree. Today we know that the triangle was too small to detect deviations

An Euclidean world could as the plane be infinitely extended, but need not be. There are also Euclidean worlds that are finite (eg a torus, see chapter 8). By contrast, the elliptical structured world is finite, it can not be infinite. It may be unlimited, but not infinite. If you run on the spherical surface straight ahead, you will never come to an end, you'll eventually circumnavigate the ball several times, although this world is not infinitely extended.

3.5 The world of space people

In order to better understand the geometry of an elliptical universe, we reduce in thoughts the four-dimensional - not unimaginable - sphere with its three-dimensional "surface" by one dimension: So we consider a three-dimensional sphere, so as we know it. Its surface is two-dimensional, ie an area. What would be the lives of individuals who have to live on this spherical surface. Our fictional universe is therefore now only two-dimensionally as a surface of a sphere.

In this area live humans, animals and plants. They all fill out no volume, but only a surface. The creatures move in their two-dimensional world, these movements may well be subject to the laws of physics, so you can define speed and acceleration.

A consequence of the lower dimension is that humans and animals can not have a continuous digestive tract, because that would divide them smoothly into two parts. Food intake and excretions should therefore be carried out in a place in the same body cavity.

Since the space-people live on a (large) sphere of life, they seem their world like a plane with Euclidean geometry. Therefore, space-scientists believe in the Pythagorean theorem and in the same geometric laws, as we know them.

Adventurers in this world who have long journeys, experience miraculous things. After a long time they come back to the starting point of their journey. They do not know that they have circled the globe. Because they know only two spatial dimensions, the notion of a sphere is completely unknown to them

Geometrically smart people were amazed about the experience that it is possible to embark on a long journey to distant lands and to be suddenly unexpectedly at home. They note that this process also occurs when they walk along a circular arc and reach the starting point again after a round. A particularly intelligent face man - a kind of Einstein - formulates the sentence: A one-dimensional uniformly curved space - and thus he means clearly the circle - has the property that, when you take it along, you come back to the starting point. And now he performs an ingenious abstraction: When this - so he says - for one-dimensional smoothly curved spaces is established, then also for two-dimensional smoothly curved spaces. He has correctly described the fact that he can not as an area man imagine, because his imagination is of course limited to two dimensions.

After our area-Einstein received the Nobel Prize his theory is further developed by his scientific colleagues. You can find out in an extremely complicated theory that if their world is a two-dimensional curved surface, the triangle would have more than 180 degrees angle sum. They then develop a sensational cosmic theory: "We live" - they explain - "in a two-dimensional curved space, which is closed in itself." Mathematicians in the area-world develop this approach further by describing the abstract concept of a three-dimensional sphere, on whose surface they live.

If we raise in this thought our model by one dimension, we obtain our universe in an elliptical geometry. We could this universe circumnavigate. This means that if we fly with a super rocket in any direction straight ahead, we are after a probably very long time back at the starting point. Today's knowledge about the structure of the universe suggests that the universe is Euclidean. But with a low probability but it could also be elliptical.

4. THE GEOMETRY OF NATURE

> *"Clouds are not spheres, mountains are not balls, coastlines are not circles, and bark is not smooth"*
>
> *Benoit Mandelbrot*

Circles, lines and bullets are objects of the geometry, we work with them and their properties are known to us. Oddly enough, you will not find all these properties in nature, they are pure idealizations. Clouds, leaves, landscapes have structures that are objects of the chaos theory.

4.1 Euclidean geometry as idealization

If engineers and architects design and build cars, houses or roads, they use mathematical calculation methods in addition to geometric objects such as lines, circles, ellipses, triangles and planes. These objects obey geometric rules that form the Euclidean geometry in its entirety. It is the geometry that we learned in school. It contains rules such as the Pythagorean theorem or the radiation theorems.

Curiously, we find these geometric objects no anywhere in nature. There are no circles, no lines and no ellipses. But wait, you might say, does not our earth pass through the path of an ellipse and is the moving if a light beam not along a straight line? False! The orbit of a planet is distorted by the gravitational influence of distant stars, planets and moons. The ellipse is only the ideal path, but which is never respected exactly. Even light does not proceed in a straight line, but on

Where does the space end?

a slightly curved path because the space itself is curved by the influence of matter. The earth is not a sphere, but is structured like a giant potato.

The elements of geometry, how we use them, therefore are idealizations. They describe the man-made reality, a reduced reality. This reality, detected by us, describes nature only approximately. Despite the uncertainty, we can predict many things with astonishing security, such as a solar or lunar eclipse.

Figure 3: One of the many forms of snowflakes as an example of a self-similar figure

The entire horizon of clouds shows an aesthetic structure.

4.2 The Geometry of Nature

If we do not find all these objects of our geometry in nature, what are the elements of a geometry that describes the nature directly?

Let's look at the example of a tree. It has nothing straight, everything is jagged and crooked. Each leaf seems to be different, while the leaves are all similar in structure. If we look at a leaf with a microscope, we would find that the leaf-edges are similar irregular. Let's look up, we see the clouds. Every cloud is irregular in its structure. If we just take a part of the cloud in inspection, we would find similar irregularities.

Here we experience a first characteristic of the natural geometry: Geometrical structures are repeated in a similar form. The chaos researchers describe this property as self-similarity. Self-similarity we find in leaves, bark, flowers, clouds and trees. Snowflakes have a particularly aesthetic form of self-similarity (cf. Fig. 3).

4.3 The Mandelbrot Set

Self-similar objects, as it purports nature, can be constructed with the aid of computers. This found Benoit Mandelbrot from the Thomas Watson Research Center of IBM in New York. He created mathematically constructable images of natural self-similar objects which were of a breathtaking beauty, as it wrote a newspaper.

In his book "The Fractal Geometry of Nature" Mandelbrot writes: "Clouds are not spheres, mountains are not balls, coastlines are not circles, a bark is not smooth." The pictures which Mandelbrot found are in their basic structure similar to these objects.

Where does the space end?

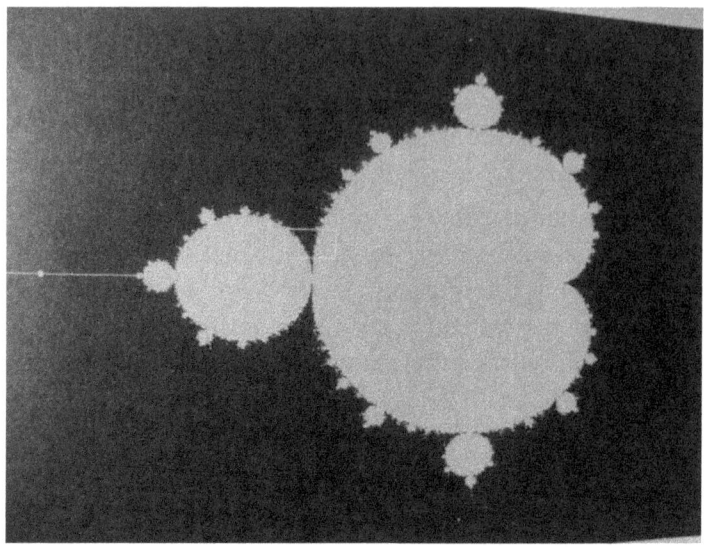

Figure 4: The Mandelbrot set

Benoit Mandelbrot, born in 1924 in Warsaw, fled from the Nazi dictatorship to Paris. There he studied aeronautical engineering, linguistics, concerned with economics. Later he moved to Thomas Watson Research Center of IBM in New York.

Mandelbrot found with the aid of complex numbers a computerized process that generates a graphical structure to the screen, the Mandelbrot set. It is called because of its shape as the "apple man" (see Fig..4)

The really interesting structures can be found in the border area between the bright and the dark area (see. Fig.4). If you would take out and zoom a tiny section from this border region, you would find images of great aesthetic (see Fig. 5). These images in turn contain other wonderful aesthetic images when enlarging small portions of them. This can go on forever and always one obtains new fantastic pictures. In addition, the images are self-similar, as the objects of nature. Mandelbrot writes in his book when he discribed the Mandelbrot set: "These images are an amazing combination of extreme simplicity and dizzying complexity"

Hartmut Jürgens, Hans Otto Peitgen and Dietmar Saupe of the University of Bremen/Germany created images of the Mandelbrot-set with high graphic devices. These they sent in two copies around the world. The exhibitions broke all visitor records of recent art exhibitions. The "Guardian" wrote: If you previously did not want to believe that in mathematics beauty might be stuck, then go in this exhibition."

Long before Mandelbrot the mathematician Gaston Julia (1893-1978) discovered these nature-similar geometric structures, when he was a prisoner of war in the hospital and dealt 1918 with studies of marginal quantities of the complex numbers. Around the same time this made the Frenchman Pierre Fatou (1878-1929). Unfortunately, both remained in theory stuck because there were no computers, with which one could visualize the structures. Therefore their work soon fell into oblivion.

The visualization of the Mandelbrot set is done through iterations in the complex plane with the aid of computers. The generation process is so simple that even any non-

Where does the space end?

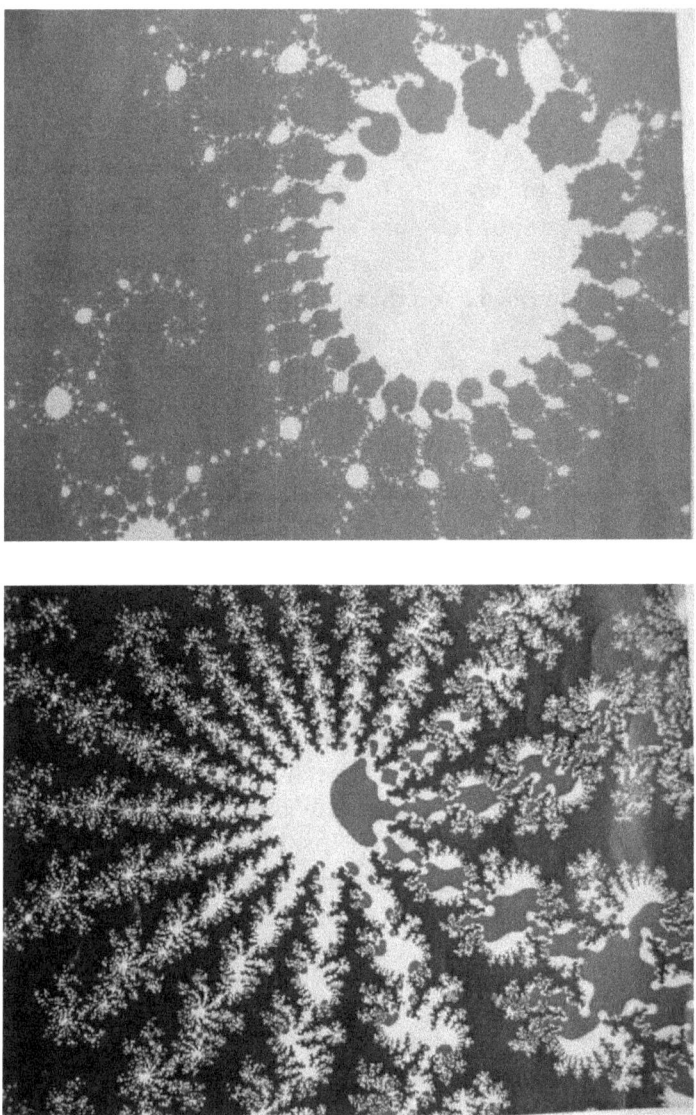

Fig. 5: Two excerpts from the Mandelbrot set

mathematician can produce Mandelbrot pictures, if he has knowledge of programming. In the appendix of this book the (easy to understand) mathematics is explained and there is given a guide for programming.

The self-similar images of Julia, Fatou and Mandelbrot apparently belong to the class of discoverable nature geometric structures.

4.4 Geometry of oddness

The forms of nature are more complex than the classical objects of the Euclidean geometry and are called fractal structure. Fractal image can be created by simple algorithms in the computer. Precursor was a self-similar curve of the Swedish mathematician Helge Koch developed in 1904 (Koch curve). This curve fits on a postage stamp, but is infinitely long.

The construction of the Koch curve is simple (see Figure 6): A straight line is divided into three parts and in the middle part a triangle is inserted (fig.6, first curve). In the second curve for each of the four individual sections the same is performed (Fig.6, second curve). When you continue this in the same manner, the curve gets always longer. A simple calculation shows:

Each curve is 4/3 times as long as the previous curve (if the increase in the middle has the shape of an equilateral triangle and all routes are of equal length). Apparently, so the curves are arbitrarily long. If the first curve in Figure 6 is ten centimetres long, then the eighth curve $10 * (4/3)^7 = 74.9$ meters long, the 15-th curve $10 * (4/3)^{14} = 561$ centimetres and 100 th curve is 233 Million kilometre long.

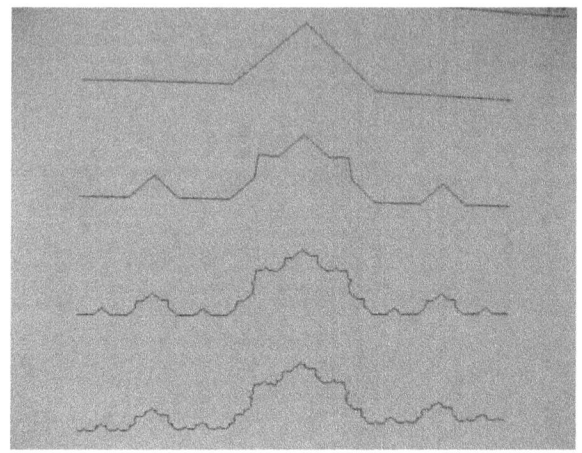

Fig.6: Koch curve

Fig.7: Sierpinski Triangle: A self-similar figure

You can go to infinity in $(4/3)^n$, then the curve is infinitely long.

Figure 7 shows another self-similar figure. The Polish mathematician Waclaw Sierpinski (1882-1969) discovered it and the figure is named after him (Sierpinski triangle). If one would look at any small part of the triangle with a magnifying glass, one would obtain again the same structure. In this sense, the figure is self-similar and this at any smallness.

The construction of the Sierpinski triangle takes place in the steps of:

1. Draw a triangle
2. Connect the centre points of the three sides.
   By this the original triangle is divided into four sub-triangles.
4. Remove the middle part of the four triangles,
   the other three triangles remain
5. Apply step 2 and 3 on the remaining three triangles.

                      etc

Circles, lines and ellipses are objects of mathematics, they do not exist in its pure form in nature. From hence we describe nature with idealized images.

5. Einstein's space-time

So what is the time? If no one asks me, I know it. If I have to explain it to a questioner, however, I do not know.

Augustine in "Confessions", XI, 14

In 1905, Albert Einstein published in Volume 17 of the journal "Annals of Physics" a review titled "Zur Elektrodynamik bewegter Körper" (On the Electrodynamics of Moving Bodies). This paper heralded a fundamental change in the understanding of space and time. So far, the idea of space was given by Newton and no one doubted its accuracy.

Einstein presented by his publication everything upside down. The room was for him no longer absolutely like a world stage, it could be shortened, was dynamic, curved and changed with the time. Considering until that time that the time was absolute just like there is somewhere in the universe a global clock, which determines the time for all regions of the cosmos, Einstein showed that this idea is wrong. Time and space form a unit, which is like a cloth made of rubber: expandable and changeable. Today his 1905 published theory is known as the "Special Theory of Relativity", short SRT.

Einstein's theory of relativity publication was entitled: "On the Electrodynamics of Moving Bodies". What has the electrodynamics to do with space? An important key to understanding the theory of relativity are the electromagnetic waves, especially the light.

5.1 space and ether

No physical phenomenon gives so much knowledge about the structure of space as the light. Therefore we will have to deal with the theme of "light" in detail.

Had Isaac Newton believed about the year 1700 that light consists of tiny corpuscles, Thomas Young was able to prove beyond any doubt at the beginning of the 19th century that light consists of waves. He observed that in superposition of different luminous flux patterns emerge that can be observed in water waves. Physicists call this phenomenon interference. So, he reasoned, light must be made of waves.

The idea of light as a wave motion was so unusual that the "Edinburgh Review" in 1803 wrote that Thomas Young's theory was directed against scientific progress. However, interference phenomena are such a clear indication of waves that the wave theory of light, finally prevailed.

But soon everybody had a problem: water waves move in the water and sound waves in the air. So waves require a medium in which they can propagate. What is the medium of light waves? Waves without medium are unthinkable. So there had to exist a medium for light waves.

Even in ancient times it was believed that the vacuum may be filled with a substance, which was called ether. The idea went back to Aristotle. In 1690 the physicist Christiaan Huygens believed that this ether, which as a fine substance fulfills the whole universe, is the carrier of light waves. As Newton later rejected the wave theory of light, this idea got into oblivion. After Thomas Young's discovery the ether theory was currently playing.

Of course, the physicists were burning keen to prove the existence of the ether in experiments, but at first without success. In 1880, the American physics professor and Nobel laureate Albert Abraham Michelson had a brilliant idea for the detection of the light ether.

In order to understand the Michelson experiment, we must go back a bit. Suppose you want to fly with a small plane from Cologne to Frankfurt. Flight Time: one hour. Once you reach Frankfurt, reverse and fly back. Apparently you're after exactly two hours in Cologne again.

What will happen when from Cologne to Frankfurt a wind is blowing at about 20 km/h? On the way you fly faster because you take advantage of the tailwind. The return trip takes a little longer because you have a headwind. First assumption: Both level out and you are after exactly two hours in Cologne.

A more accurate statement shows, however, that this is not like that. You need about 5 minutes longer for the whole flight. Michelson concluded that also the following must apply: Our Earth moves at about 29.78 kilometres per second through space, so also through the ether. Just as a motorcyclist at high speed feels headwind even without wind, so it must be some kind of ether wind on Earth. So a ray of light, if you send him against the ether wind in the direction of movement of the earth and then by mirroring back, has a longer duration than a beam which is sent perpendicular to the earth-motion and thus there is no ether wind. So Michelson sent in both directions light rays which again came back each by mirroring.

Using the finest measurements (interference) Michelson found no runtime difference.

He doubted his experimental arrangement and repeated the experiment with the American chemist E.W. Morley, but with the same result. This experiment is called as "Michelson experiment" in the physics history, and was repeated by many researchers with higher accuracy. No one could notice a runtime difference.

There were erected many theories to explain the experimental result, but no theory could convince. Around 1900 there was a physicist of the Federal Patent Office in Bern who based on the simple assumption that there might be no ether. His name was Albert Einstein. This would explain the outcome of the Michelson experiment and it was the birth of the theory of relativity. Einstein's ideas were based on publications of the Dutch mathematician and physicist Hendrik Lorentz, who had developed between 1892-1904 formulas, in which the speed of light is always the same and constant, no matter what direction the light beam moves and from where it is emitted.

The oddity of this statement, we will look in more detail in the next chapter. It became the basis of the special theory of relativity and the consequences in terms of space and time were revolutionary.

5.2 The speed of light is constant
What the notions of space and time changed so radically was the knowledge of the speed of light.

The measurement of the speed of light had been performed already at 1675 by Olaf Römer by evaluating the orbital periods of the innermost Jupiter moon.

The measurement methods were refined more and more over time. The calculated value of Olaf Römer departed only 30% of from the value known today. One usually uses the symbol c and it is now: c = 299 792,458 km per second. Simplified we can say without making a big mistake that c = 300.000 km per second. At this speed, a ray of light circles the earth more than seven times per second.

The absence of an ether and the formulas of Lorentz showed that the speed of light is always the value c = 299 792.458 kilometres per second, regardless of where and in which direction the light is emitted. This is known as the constancy of the speed of light.

The constancy of the speed of light represents our ideas about space and time on the head. This shows the following example:

You drive on the highway with your car with constant 80 miles/h. On the left lane overtakes a car at the speed of 120 miles/h. Then, the differential speed is 40 miles/h that means the overtaking car moves away from you at 40 miles/h. The reason: The cars move relative to each other.

With light it is different: If a ray of light at the speed of c = 299 792.458 kilometres per second "overtakes" you, the difference in speed to your car continues to c = 299792.458 kilometres per second, and no less.

The oddity of this statement is obvious, if we accept the thought experiment, that the speed of the second car applies the constancy of the speed of the car as it is with the light. In this case you overtake the car at 120 miles/h, but the speed difference to you after overtaking is further 120 miles/h. This is because the "constancy of the speed" means: The speed is always the same, no matter where and how I measure. So it is with light..

5.3 speeds alter space and time

This is beyond all our ideas about time and space. How can that be? If the speedometer of the overtaking vehicle is actually 120 miles/h and your speedometer is 80 miles /h, then the speed difference can not be 120. miles / h.

There is only one way out of this dilemma: One of the two tachos or even both are incorrect in measurement. Since the speed is always the distance travelled divided by the time, it could be, for example, that the overtaking car and its speedometer perceive space and time differently.

In fact: Formulas and experiments of physics show that the time passes more slowly in a moving system like a rocket or an airplane. So there is in the universe no super-clock that purports the time to all areas, as Newton still accepted, the time is relative. That is the basic assumption of the theory of special relativity. So we formulate the important law:

In a moving system the time passes slower

The implications are astounding: When you run, your watch (or more precisely your time) is actually slower than when you are standing. However, at these low speeds, the time differences are so small that they are not measurable. At high speeds, such as approximately half the speed of light, these differences can be substantial: at half the speed of light, a minute shortened to about 52 seconds (see Table 1). In the navigation system GPS, which is controlled by satellite, the delay difference must be considered because of the high velocity of the satellites.

Just how much the time slows down is shown in the following table (time dilatation):

Rocket speed in percent of the speed of light	a minute on Earth corresponds to the rocket-time in seconds:
10%	59,7
20%	58,8
30%	57,2
40%	55,0
50%	51,6
60%	48,0
70%	42,8
80%	36,0
90%	26,2
99%	8.5
99,99%	0.8
100%	0

Table 1: The time dilatation.

5.4 The space can be bent

The European Space Agency has decided in 2012 to send in ten years the space shuttle "Juice" (Jupiter Icy Moon's explorer) to the planet Jupiter to explore the three Jupiter moons Ganymede, Europa and Callisto.

The space shuttle will have to overcome a distance of 800 million kilometers and will be on the way for eight years. Would astronauts be on board, they needed including return therefore 16 years. But if the time may be reduced at high

Where does the space end?

speeds, you could send a spacecraft that moves at half the speed of light. Then the traveling-time would be shortened.

But so high speeds can not be realized. There are many reasons which will not be discussed here. But we can imagine in a thought experiment that such a trip could run at such high speeds.

So we assume an astronaut team starts to Jupiter, 800 million kilometres away. The spacecraft flies with 222222 kilometres per second, which are 800 million kilometres per hour. But when it travels 800 million kilometres in one hour, our spaceship reaches the planet Jupiter after one hour.

Compared with the time on earth the matter presents itself differently. Because of its high speed the spacecraft has an another time. The time is slower. The trip did not take an hour, but only about 40 minutes, as can easily be seen by Table 1. According to their own board protocols the astronauts fly at the speed of 222222 kilometres per second and after 40-minutes they reach Jupiter.

We make the following calculation: 40 minutes are 2400 seconds. For 2400 seconds they fly at a speed of 222222 kilometres per second. This makes 2400 * 222222 = 533 million kilometres. Jupiter is 800 million km away but. Only 533 million kilometres flown and already arrived? There is only one logical conclusion: not only the time has shortened, but also the distance.

This is in fact the statement of the theory of relativity of Albert Einstein. According to Einstein's theory there changed at high speeds not only the time but also the distance. The distance becomes shorter. The distance to Jupiter is actually for travelling astronauts during their high speed - seen from us - just 533 million kilometres. That time and space are shortened is not noticed by the astronauts. The changes are made in terms of observations from Earth.

Another example: A car is driving at the speed of v = 0.8 * c, (0.8 times the speed of light). It passes an observer on the roadside. In the car is a rod of the length 30 centimetres. For the observer on the roadside, the car is moving, so the rod is shortened. The account of the theory of relativity gives a length of 18 centimetres. The rod may be thrown out of the car on the road. If the observer now again measures the length of the rod, it is 30 centimetres long, because for the observer the rod is now at rest. However, the length from the perspective of the car-driver is now 18 centimetres because the rode moves relative to the driver. Thus, the length contraction turns out symmetrically.

So we hold: _

> *In a moving system the space*
> *is shortened in the direction of movement*

5.5 The Searchlight effect

Could we drive through an alley at a rate that is close to the speed of light, we would perceive the space curved. The curvature we would notice on the houses, as the picture 9 shows. However, here we have no relativistic deformation of the space.

Where does the space end?

Fig.8: Could a car with half the speed of light travel, it is shortened to an outsider as it shows the right picture. The driver does not notice the shortening.

The reason of the apparent deformation is that light beams emitted from the tops of houses, need a slightly longer route than the rays emitted from houses in eye level.

So they were a little earlier emitted when they reach me (tiny fractions of milliseconds). The result is a perspective shift which causes the apparent curvature. This phenomenon is referred to as a searchlight effect.

Where does the space end?

Fig.9: Would you move at high speed through this alley, space and houses would appear curved

6. CURVED SPACES

*Since the mathematicians attacked
Relativity I understand it myself no longer.*

Albert Einstein

Space and time vary at high speeds. This was the result of Einstein's special theory of relativity. Ten years later, Einstein completed his theory of general relativity, which asserts that the gravitation changes space and time.

6.1 Gravitation

Why does the moon revolve around the earth and does not escape into space? Newton postulated a force that connects earth and moon together as a band. All the stars and planets exert an attraction on the neighbouring stars and on the neighbouring planets. This force (gravitation) holds the universe together and is universal, static and can be represented in equations. The equations of Newton permit a (almost) correct calculation of the orbits of planets and satellites.

Albert Einstein asked himself at the beginning of the 20th century how such a force can come. He found a revolutionary solution: The cause of gravity is the space. The room is not like Newton a static and unchanging frame, a kind of stage for world affairs, but dynamic, flexible and changeable. The cause of the gravitation is the curvature in the space.

This curvature is inconceivable, but available for us. But we can make it visible when we look at it not in three but in two dimensions. We then have a surface which is curved and which you can imagine. The easiest way is to take a blanket, similar to a trampoline. In the centre we place a heavy weight. The cloth bends in the middle at the bottom and the surface of the sheet is curved. Would we attach to the edge of the cloth a tennis ball it would roll into the centre toward the heavy weight in the middle. In exactly the same way the moon is attracted by the Earth: The Earth is in this picture the weight in the middle and the moon is the tennis ball. The curvature of space causes the attraction.

The deeper bulge of the blanket, the heavier the weight in the middle. Transferring this means: the more mass a star has, the deeper is the bulge of the room, the stronger the curvature of space around him.

Although we live in a three-dimensional space, but can not imagine its curvature. However, Albert Einstein delivered complicated equations that permit the calculation of the curvature. There are the equations of General Relativity, which he published in 1916. They show clearly that the curvature is caused by matter and energy. Where no matter and energy, there is also no curvature. If the density of matter or energy is high, the curvature is pronounced. Particularly messy is the curvature near burned-out stars, the black holes. Since space and time form a unit, the physicists speak of the curvature of the spacetime.

Where does the space end?

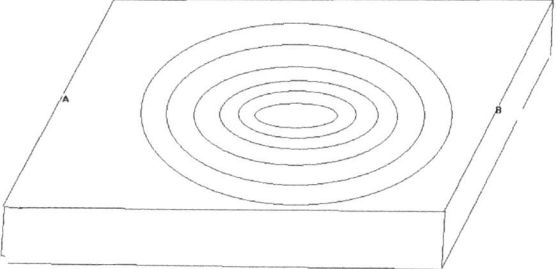

Fig.10: The above surface is curved and has a depression in the middle, also a three-dimensional space may be curved. A ball that starts at the edge, ends up in the dump.

$$R_{\mu\nu} - 0.5 * R * g_{\mu\nu} - \Lambda * g_{\mu\nu} = 8 * \pi * G / c^4 * T_{\mu\nu}$$

The equations of Einstein to describe the curvature of space

6.2 Lengths are shortened

Let's get mass into the vicinity of a star, then this mass is attracted to the star.

We envision a metal ball in the vicinity of the earth. The gravity causes a force with which it is drawn down to the earth. Let's go with the metal ball to great height, the power is low, because the gravity is lower then. The higher we go, the smaller is the force of the gravity. We could now measure and record the force at any point outside the Earth. Then we get what the physicists call a gravitational field. Each star is surrounded by a gravitational field

If we bring an elongated mass (eg, a rod) in a gravitational field, the mass (rod) is shortened, namely in the direction of attraction.

As we saw earlier, lengths are shortened at high speeds. Now we can supplement this statement by stating that a reduction takes also place in gravitational fields. The space deforms in a gravitational fields always in direction of gravity acting. The shortening is a consequence of the curvature of space in the gravitational field. This say exactly the equations of general relativity, as Einstein formulated.

6.3 Experiments

Is Einstein's theoretically predicted curvature experimentally justified? Immediately after Einstein's publication the Austrians Josef Lensing and Hans Thirring showed that a rotating body drags the curved spacetime around it and "twists" it (Lense-Thirring precession).

Leonhard Schiff and William Fairbank Sr proposed in 1962 an experiment, which is associated with the name Gravity

Probe B (GP-B) and builds on the Lensing-Thirring effect. Four fast rotating quartz balls were floating freely on the Satellite GP-B and rotated in vacuum at 10,000 revolutions per minute. If the space is curved near the Earth, the axes of rotation per year would have to change to a tiny value. On 20 April 2004, the satellite was launched successfully. The measurements proved more difficult than expected and the project had to be extended, resulting in high costs. The project was interrupted nine times. 2008 appeared the first final report, the final appeared 2011. The experiment confirmed with an uncertainty of 5% the Lense-Thirring effect.

6.4 Gravitation changes the time

As illustrated, the speed of light is always the same, regardless from where the light is observed. It has always a value of about 300,000 kilometres per second. We assume that a rocket gets into a gravitational field and flies in the direction of gravity. The space is shortened in the direction of flight and we further assume that gravity is so strong that the space distance is reduced to a half. 300000 kilometres then are reduced to 150000 kilometres. Was the speed of light without gravitational 300000 kilometres per second, it would now be for the astronauts in the rocket only 150000 kilometres per second.

This can not be, because even inside the rocket the speed of light must be 300000 kilometres because of the constancy of the light. We will solve the problem in the way that we assume that the time also slows down, exactly to a half. If a second is reduced to half a second, then the speed of light is now: 150000 km divided by half a second, which is 300000 km per second. And the physics of Einstein is back on track.

Where does the space end?

In fact, it is so that in a gravitational field the time goes slower than in gravity-free space. The stronger the attraction of a star, the slower goes the time.

In 1959 the time delay (the physicists speak of time dilatation) was conformed at Harvard University experimentally. On a high tower the earth's gravity is somewhat less than at the bottom. So there should extend the time a little faster. Using nuclear timekeeping time was measured on a 22.5 meter high tower and compared with the time on the ground. It was actually a difference. This, however, was tiny because of the low height. The time difference was only 0.000000000000257 percent. This is still only 0.0008 seconds in a hundred years

When the time runs slower, then biological processes proceed more slowly and thus aging is more slowly. In this sense, high-rise residents age faster. But they lose against citizens on the ground floor - while a life - only fractions of a second.

More the time delay affects satellites orbiting the earth at a high altitude. Since the Earth's gravity at this level is substantially lower than on the Earth, the satellite clocks go faster. On the other hand however, the satellite moves at a high speed and - as we saw - moving clocks go slower. Is the satellite clock now slower or faster?

Which of the two effects predominates, it depends on the height of the satellite. At an altitude of 3200 kilometres, the two effects cancel out. Both constructions for navigating satellites must be considered.

7. THE UNIVERSE

> *People amaze me who want to understand then universe where it is difficult enough is to cope in Chinatown.*
>
> *Woody Allen*

Our habitat is the Earth. The vast oceans raise up the impression that the space where we can travel is huge. And yet the earth is only a small celestial body, compared to the size of the sun. The sun holds eight planets and forms the solar system with them. The solar system, in turn, is tiny compared to our home galaxy, the Milky Way, which contains billions of suns.

7.1 The Solar System

We, as the inhabitants of the planet Earth travel with an average speed of 29.78 kilometres per second, so with about 107200 kilometres per hour or 2.57 million kilometres per day through the room. Our earth has to move in its orbit around the sun so fast that it creates a circumnavigation per year. Would it be more slowly, the earth would soon drift towards the sun, and if it would fly faster, it would escape into space..

Our earth is tiny, when compared to the size of the sun . Had the sun the size of a pumpkin, the earth had only the size of a pea, which orbits at a distance of fifty meters to the sun (see fig.11).

Fig: 11: The world is as big as the point in the picture above, compared to the sun. In this figure, the Earth would be (the point) located at 6 meters from the drawn sun

In addition to the Earth, the Sun has eight planets: Mercury, Venus, Mars, Jupiter, Saturn, Uranus and Neptune. Pluto was stripped of 2006 the status of a planet and relegated to the category of dwarf planets. They all form a system with an expansion radius of nearly five billion kilometers.

Climate and nature of the planets have on Earth ideal conditions for the development of life. This appears not to be the case for the rest of the planets.

For a long time Mars was a candidate for life. However the space probes that landed on Mars founded so far no real traces. They founded carbon, but this may also have a different origin than life molecules. The surface of the Mars consists almost entirely of desert, and the average temperature is minus 63 degrees Celsius.

The nearest planets to earth are Mars and Venus. On June 6, 2012 millions star-enthusiastic people stared with sunscreens on the sun. A small dark visible spot passed the sun. It was the planet Venus. This spectacle will only be back in 2117.

The neighboring planet Venus is the planet that comes pretty close in mass, density and size up our earth. A day on Venus lasts 243 Earth days, the Venus year is shorter than a day on Venus. New Year and New Year's Eve can therefore fall on the same day. The sunrise takes months, daylight is very sparse, there is a thick cloud swallowing the light. Since Venus is closer to the sun than to the Earth, the temperature is up to 500 degrees. At the same time there is a pressure of 90 atmospheres.

The other planets and their moons are similarly inhospitable. There are cyclones with wind speeds of over 500 kilometers per hour and rivers where no water flows, but methane, which is liquid at the local temperatures.

The vast emptiness of the universe

The sun and its planets and their moons are located in the

Milky Way galaxy. This is a star island in space and is home to about 100 to 200 billion suns. To traverse the Milky Way in the longest dimension, light takes 100,000 years. A beam of light which was launched from the edge of the Milky Way 2000 years ago and should fly through the maximum extent of the galaxy would have managed just 2 percent of the distance to flying through until today. Could we fly through this route with an Airbus A380, we need 120 billion years. That's about nine times the age of the universe.

The Milky Way has (see fig.12) the shape of a spiral. Like the earth rotates around the sun, so our galaxy rotates around its centre. A full rotation takes around 230 million years (a galactic year). Our solar system is moving at 230 kilometres per second around the centre.

To our galaxy around is a huge empty space. The nearest large galaxy, the Andromeda galaxy, is 2.5 million light years away. The light requires 2.5 million years to reach us. Between the two galaxies - apart from a small dwarf galaxy – is an empty space.

We start for a reconnaissance trip into space. The speed of a normal spaceship would be much too low to get very far. We choose the highest speed that exists, namely the speed of light with 300,000 kilometres per second. Of course, we will never be able to build spaceships with this speed, but all that should be in our thought experiment neglected.

In a thought experiment we start on Earth. Already after eight minutes we pass the sun and after four or five hours we leave the solar system, as we fly past the outermost planet Neptune.

We now fly days, months and even years to pass without any star. Four long years, we are on the road, until we finally reach the first star, namely Proxima Centauri. Proxima

Centauri is the star that is closest to the Sun, is less than the sun and can be observed only from outside Europe.

Fig.12: Spiral Galaxy

After 160,000 years, we pass a small neighbouring galaxy, the Magellanic Cloud. After that we go millions of years without seeing stars or galaxies. Only after 2.5 million years we reach the Andromeda galaxy, the nearest large galaxy.

The Andromeda Galaxy is larger than the Milky Way. Both galaxies rush toward each other with a single track approach-speed of 150 kilometers per second. Eventually it will come to the big crash. Then Milky Way and Andromeda will unite into a giant galaxy.

We leave the Andromeda nebula and fly for long periods only in empty space. It may happen that we fly many millions of years with the speed of light, without encountering any matter or galaxies. We lose ourselves in the eternal darkness of the cosmos. Sometimes we can see in the distance strange luminous formations with telescopes: the galaxies.

The distribution of galaxies in space is similar to large scales of a honeycomb structure formed by filaments with huge voids in between. These voids denote astronomers as voids. Voids usually have a diameter of 100 million light years. In the constellation Eridanus was discovered, a void having a diameter of about one billion light-years.

To get an overview of the proportions in space, we reduce this in scale 1: 1 billion. If we leave the time unchanged, a light year would now have the distance of less than 10000 km and the speed of light would be about 30 centimetres per second, which is 1080 meters per hour. The sun is now a fireball of about 1.40 meters in diameter, surrounded by earth at a distance of 150 meters. The Earth is only 1.2 centimetres in diameter.

The outermost planet Neptune has a distance of 4.5 kilometres. Behind comes a huge empty space, because the next shining star is nearly 40,000 kilometres away. Would be at this scale our solar system somewhere in Europe, the nearest star would be twice as far away as Australia. Up to a distance of 140 000 kilometres, there are only 20 more luminous stars.

7.3 The universe from the perspective of the astronauts

In our journey through the universe we could fly millions of millions of years and would go again and again into new regions. The size is unimaginable in its expansion.

Consider, however, the expansion of space from the perspective of a spaceman, who crosses the room with almost the speed of light, you would have a totally different picture. The faster it flies, the smaller the space. Time and space are reduced at high speeds so that the orders of magnitude look different.

Suppose we enter our space journey on a spaceship which flies through the space with almost the speed of light. Of course, this will never be possible, because at such high speeds, the mass of spaceship and astronaut are so great that no energy is sufficient to bring the necessary acceleration, but this we neglect in the following thought experiments.

We must differ between the board time of the spaceship, that is the time that displays the clock in the spaceship, and the earth time, ie the time it takes for an Earth observer. The faster the astronaut flies, the slower is his clock and the shorter will be the distances (see chapt. 5.3).

At 90% of the speed of light, the flight time reduces to 26% of the earth-time, if the time is measured from the Earth. In 99.99% of the speed of light it reduces even to 0.8%. Accordingly, the distances that have to be overcome are shortened.

.

7.4 Big Bang and expansion of space
The universe has a beginning. 13.8 billion years ago there was a huge explosion, the Big Bang.

When there are explosions, matter is ejected with high energy into the surrounding space. At the Big Bang it was different.

Where does the space end?

The surrounding area was not yet available. Space and time emerged simultaneously. Not matter flew into the room, but the space evolved from a point out and then expanded at high speed. Not individual chunks of matter were flying apart, but the room was getting bigger. The expansion of space and time began its course.

What was before the Big Bang? A "before" implies the time, but time did not exist before. The medieval philosopher and theologian Augustine was asked: What did God do before he created the universe? Augustine replied jokingly: He created hell for those who ask such questions.

It was not until the time of

0.0000000000000000000000000000000000000000001

Seconds (ie, 10^{-43} seconds), the laws of nature were known to us. This time is called the Planck time. What was before the Planck time and what was the cause of the Big Bang, this is unknown.

The mass density of the then tiny universe was 10^{94} grams per cubic centimetre, which is a number with 95 digits. The temperature was 10^{32} degrees Kelvin.

The universe expanded. Within the first second, the universe enlarged according to the experts by a factor of 10^{43}, this is a number with 44 digits. This enlargement, called inflation, came in a tiny fraction of a second. Then it followed a normal expansion that continues to this day. The universe expands. A 100,000 light-years distant galaxy is moving away at 2.3 kilometres per second, or about 8300 kilometres per hour from us, one light year is the distance that light travels in one year. If we regard a Galaxy three million light years away, it has the escape velocity of about 70 kilometres per second. The rate of expansion is greater the farther the galaxies or stars are removed.

The expansion is comparable to the increase in the surface of a balloon as it is inflated. We provide the balloon with spots on its surface, so the points are further and further away from each other, if you inflate the balloon. Just so does the expansion of the universe. This is evident, for example, in the fact that the wavelengths of light itself enlarge, thus expanding

If one compresses air, it heats up. Every biker knows that his air pump warms up when pumping. Reversed: Stretches from air, it cools.

Applied to the universe this means: Stretches the All out, it cools off. One second after the Big Bang, the temperature was still 10^{10} K or 10 billion degrees Kelvin, after 370,000 years "only" 3000 K. To this day, the universe has cooled to the temperature of 2.7 degrees K.

7.5 Stars and Galaxies

At first elementary particles, including electrons and protons raced through space. As the temperature dropped because of the expansion of the universe, the electrons were slow and were able to combine with protons. The first light atoms such as hydrogen and helium were created. This happened about 370,000 years after the Big Bang.

Now the universe was filled with a gas mixture consisting of hydrogen and a lot of little helium. In some places, the density of the gas was slightly higher than average. In these areas matter accumulated. By gravitation here the gas compressed to clouds and inside the clouds there was a high pressure. After hundreds of millions of years here arose the first galaxies and galaxy clusters and in these arose the first stars. A galaxy possesses 10^6-10^{12} stars. Many of the stars are thousands to millions of times heavier than the sun. The largest ever discovered star is one billion times larger than our sun. If the stars were as small as tennis balls, one could fill with the "tennis balls" of one galaxy several stadiums. The galaxies enforce the universe as islands in the sea. Our solar system is located just outside the centre of the galaxy "Milky Way". The light requires 100,000 years to traverse the Milky Way.

A star is formed by first burning of hydrogen into helium. This results in enormous amounts of energy that is radiated into space as light and heat. Our sun is currently at this stage.

What happens in the sun when all hydrogen is burned? The sun is then mainly of helium and this substance burns now to carbon. Could we observe the process from the earth, we would see that the sun is getting bigger at this stage and the light changes into the red colour. The sun would eventually cover the entire horizon, the sea would dry up and in the final phase the sun would be so large that it would swallow the whole Earth. The cosmologists speak of a "red giant". At the end of the sun remains a small remnant star, called a white dwarf. It is the ashes of all burns in the sun

7.8 Supernovae
Star of the order of the sun end as a white dwarf. Slightly larger stars may be the "planetary nebulae". This has nothing to do with planets. Around the core is formed a gas cloud

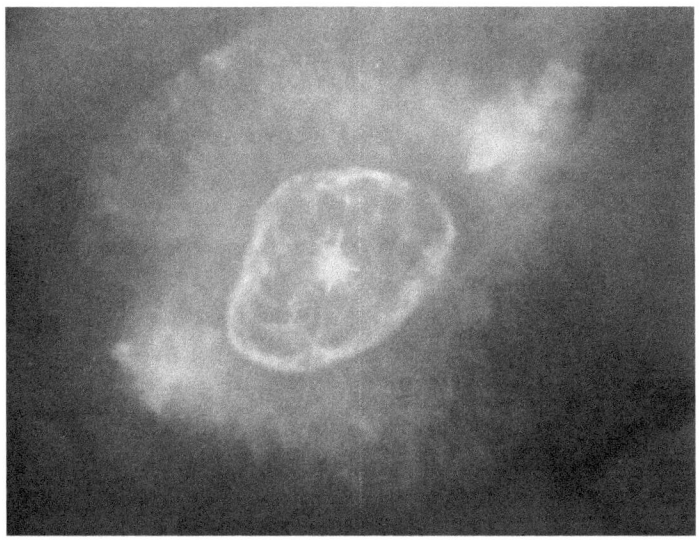

Fig.13: The planetary nebula "Cat`s eye '

consisting of hydrogen, helium and other materials, which has a width of one light year and spreads slowly. The gas cloud shines in different colours from red to blue (see. Fig.13). In our home galaxy are about 1500 planetary nebulae.

The death of a star that is at least nine times heavier than the sun, takes place under dramatic circumstances. It explodes with unimaginable violence and with immense energy. The luminosity of the explosion can be up to billions of times brighter than the sun. This is a supernova. On February 23, 1987, scientists observed a supernova in our neighbouring galaxy, the Large Magellanic Cloud. The explosion took place in 165000 light years away.

What happens when a supernova explodes? After hydrogen and helium, the lightest elements, are burned to carbon, silicon arises, neon and oxygen. There arise fusion processes that take approximately a year and release a lot of energy. Therefore, the star is getting hotter. After that arise nickel, cobalt and iron within a day. The combustion phases are becoming ever shorter and release enormous energies. The star now lights up with a luminosity that is many thousands of times greater than that of the sun.

In all fusion processes energy is given free, but only until inside only iron is produced. When this is the case, the burns stop. Now no more fusion energy is produced. One would think that now the star comes to rest, but the opposite is the case. From now on it is dramatic: The Burns had generated a pressure pushing against gravity, thereby preventing a collapse of the star. This pressure is now eliminated.

Imagine a cage with wild animals. The animals jump against the cage wall, race around and want to leave the cage. They practice with their movements pressuring against the cage walls. If they get tired after some time, it diminishes the pressure on the cage walls.

The pressure inside the star behaves like that. It was initially against gravity. If the pressure is omitted then the gravity has free rein. All atoms of the star rush simultaneously to the core, so with tremendous speed. This results in new elements such as gold and uranium. When they reach the core, they are like tennis balls that bounce on a wall thrown back. A shock wave hurls the atoms at a rate of 1000 to 20000 kilometres per second outwards into the surrounding space. An immense explosion, called a supernova. Many of the chemical elements in the planet formed in a supernova. The supernova is the furnace in which the elements are "cooked" , from which we are made.

A special drama is produced in a binary star system in which one of them is a white dwarf. From the companion material flows over onto the white dwarf and this grows and becomes larger. Its temperature increases to over a billion degrees and the density reaches 1000 tonnes per cubic meter. Eventually, it tears the formerly white dwarf and its matter is racing with a few thousand kilometres per second into space. Its brightness is

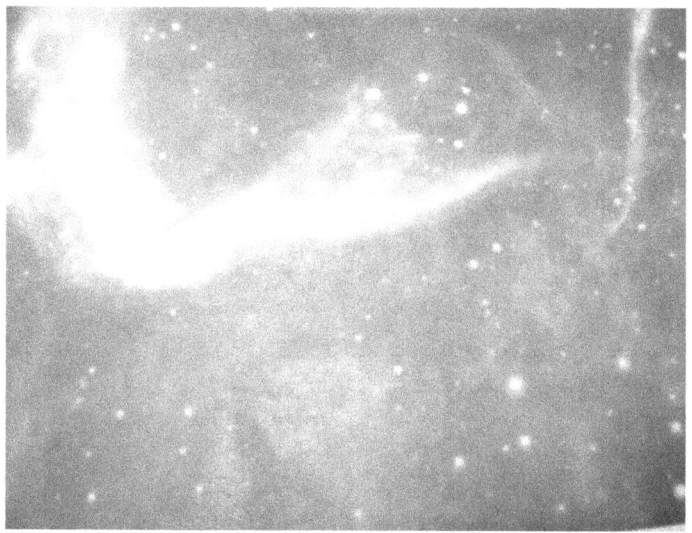

Fig.14: remnants of a stellar explosion

now up to 10 billion times brighter than the sun and the luminosity can accept the luminosity of an entire galaxy ..

7.9 Are there aliens?
A private television station in Germany announced a science program in July 2013 with the sentence: "Scientists largely agree that there is extraterrestrial life." In the same year the

planet researcher Sara Seager of MIT told in a conference at Harvard University that it would take another 10 years until finally planets were found on which sprouts life.

So far the search for aliens or extraterrestrial life, however, was not very successful, although a handful of scientists and millions of amateur astronomers are looking tirelessly for it. Several times, NASA announced that traces of life were found on Mars, but they always had to withdraw this later. Serious astronomers are reluctant in this matter and prefer to let speak their instruments.

What was the reason for the optimism of Mrs Seager? A group of astronomers had discovered in the neighbourhood of 22 light-years distant dwarf star Gliese 667C three planets on which there could be liquid water. Water is known to be the basis for life, so there could actually be live there, so Mrs Seager. It is based on a formula that goes back to the original version to the astronomer Frank Drake in 1961. In this formula Drake multiplied seven probabilities together and came as the probability of extrasolar life. However, Drake failed miserably with this formula. Mrs Seager interpreted the formula new on the basis of recent observations.

There is no doubt so that the probability of the existence of Earth-like planets is very high. However, the question of the origin of life on this planet is completely excluded. The probability of a first cell capable of division is formed by chance somewhere in space, is almost zero. The astronomer and evolutionary scientists Fred Hoyle compared the random emergence of a purely first divisible biological cell with a waste disposal site, lying scattered on it all the items of a Boeing 747. An incipient hurricane swirls these parts then randomly so messed up that after the storm in the square a flight Boeing is ready. This means that in his opinion the random genesis of a cell capable of dividing is similar, is virtually nil.

One can therefore hold: The probability that Earth-like extrasolar planets exist is extremely high, the probability that there is created life by chance is almost zero. If we combine both statements, it is as if we would infinitely multiply by zero (which is forbidden in mathematics). It may all come out. So it is with the question of extrasolar life: It can be anything possible. Life could exist, but that it may also be that the earth is the only planet which carries the life.. That - according to statement of the aforementioned TV station - the scientists were largely in agreement that there is extrasolar life in space, is a wishful thinking and has nothing to do with reality.

7.10 Dark Matter

Our home galaxy Milky Way has the shape of a disc that rotates. Imagine a rotating carousel. Would it turn more quickly, the passengers would be thrown at some point outwards. It is similar with the rotating Milky Way: Stars on the very edge should be thrown outwards and leave the galaxy. Instead the outer stars move more slowly than the stars in the centre.

Amazed it was discovered a few decades ago that this is not the case. The rotational speed (more precisely, the angular velocity) is in the same size as in the centre. Nevertheless, the outer stars remain in its orbit, although the calculations yielded different results.

It must be therefore in addition to the normal gravity another force that keeps the outer stars. One possibility is that there is not visible matter, which by their attraction (gravitation) prevents outer stars to be thrown out. It originated the term "dark matter", which is thought to be present in the interior of the galaxy.

There are other indications of dark matter: The universe would fly apart if it were not held together by dark matter. This show at least computer simulations.

How much dark matter is in the universe? From the movements of the stars and from other sources it could be calculated the share of space matter. Then it must be six to seven times as much dark matter than visible matter. The substance of the stars that we see is only a small part of what is present in the universe. The larger part is dark, that is, not visible.

What is dark matter? The answer is simple: We do not know. Consist the visible matter of atoms, and these in turn from elementary particles, this does not seem to apply to the dark matter. It sends, for example, no radiation, as is the case with the atomic visible matter. Therefore, it is invisible.

7.11 Dark Energy

Since almost 100 years it is known that the universe is expanding. It was assumed that the expansion would continue forever or would eventually turn into an implosion.

Two groups of researchers attempted in the 90's of the last century to figure out how fast the expansion proceeds. To their surprise and to the surprise of their colleagues, they discovered that the universe is expanding faster and faster. The rate of expansion seems constantly increasing. In 2011 they were awarded by the Nobel Prize in Physics for their discovery.

The previous knowledge of cosmology provides no

explanation for this state of affairs. If the rate of expansion increases steadily, there must be responsible any energy. But which?

Until today this question is unclear. This unknown energy is called dark energy. The proportion of dark energy in the total energy can be calculated. The result is astonishing: In the energy budget of the universe the dark energy makes about 73 percent. If we take into account the high proportion of dark matter, is left only 4% for the visible and understood matter. 96% of the available energy in the universe is absolutely unknown to us. We only can speculate.

Two thirds of the earth's surface is covered by water and it would be strange if we did not know at this constellation, which is water. We can describe only 4% scientifically about the energy in the universe, the rest remains unknown.

In Chapter 2, we dealt with the vacuum energy: Each vacuum is filled with energy. Could the vacuum energy be the key for us to understand? Attempts have been made to clarify this issue in the corresponding invoices. The result: If the vacuum energy is the trigger for the accelerated expansion, the energy supplied would be a billion times greater than it is in reality. The question of the cause of dark energy and dark matter therefore remains unclear.

Albert Einstein predicted the dark energy unknowingly, but later he recanted his testimony. At the time, the expansion of the universe was not yet discovered, people believed in a static universe. The equations of general relativity, however, gave an expanding universe. Thus Einstein introduced into these equations an additional constant, which became known as the cosmological constant and it did not falsify the statements of equations. Einstein chose this constant so refined that a possible expansion could no longer take place. His universe was static. When Edwin Hubble later discovered the expansion, Einstein recanted his correction, calling it "as the greatest error of my life". Today we know that the cosmological constant is suitable for describing the cosmic expansion when it is chosen suitable.

7.12 Multiverses

You hear it more and more: There are many other universes, many without any life, as it were a stillbirth. The reason: The laws of cosmology and the laws of general relativity allow many, even an infinite number of universes. They speak of a multiverse. Our universe is only one of them, in which randomly created life. There are scientists who deal with the speculative ideas in monographs or journal articles, thereby consciously or unconsciously creating the impression as if the existence of Mulitiverses is as good as proved.

Is there any truth to these statements? There is not a single proof of the existence of a multiverse. Also an experimental verification will never be possible in principle. That the equations of relativity theory allow several universes, does not mean that they actually exist. Not everything that mathematics allows is real. As a simple example: consider the equation of Pythagoras for a right triangle:

$$c = \sqrt{a^2 + b^2}$$

Here, c is the length of the right angle opposite side (hypotenuse). Root expressions can be negative so as

$$\sqrt{4} = +2 \text{ and } -2$$

but there are no triangle with a negative side length. The mathematics clearly describes more than there is actually present.

Is therefore the statement about multiverse pure speculation? Peter Woit, a mathematician at Columbia University in New York and author of many books, called faith in multiverses as wishful thinking. Even at the time of Albert Einstein there were scientific speculations, which lacked any serious evidence. Einstein wrote: "Whoever invents, the events of his imagination appear so necessary and natural to him that he sees them not for structures of thought, but for given realities"

Let one of the most famous cosmologists speak on this subject. The astrophysicist Brian Schmidt is one of the discoverers of the accelerated expansion of the cosmos. In 2011 he received with others the Nobel Prize. On the occasion of the Nobel Prize, he was asked in an interview after his opinion about multiverse. (FAZ, Germany, 7.12.2011). Brian described himself in this matter as "agnostic" and literally said: *"I refuse to believe in things that I can not verify. ... .Who in multiverse theories is working, but does not believe that they will ever be able to be verified, is not a scientist."*

7.13 The Evolution of the Universe

How is the development of the universe, if we compress the time since the Big Bang to the present day to one year? Then one billion years would roughly correspond to a month.

On January 1, around 0 o'clock arise in an act of creation, which we call the big bang, time, space and matter. The room is filled with a primary matter, which is afflicted with a tremendous density and with a gigantic high temperature (energy). Already in the first second from this basic energy arise elementary particles and from these the first atomic nuclei and soon after that arise light atoms such as hydrogen and helium. Already end of January the first galaxies are formed

In the hydrogen retention burns hydrogen into helium, the first luminous stars have been formed. This is the stage where our sun currently is. Later, helium burns to carbon and high-order elements. In massive supernova outbursts occur atoms from which planets are formed. This all takes place in the following months.

Our solar system with its planets is formed in mid-August. In mid-September starts life on Earth, by the formation of the first single-celled organisms. Algae, aquatic plants and aquatic species occur until mid-November. On December 2, from fishes arise the first land animals from which came reptiles. On December 20, the continents are covered with forest, it forms an oxygen-containing atmosphere. At Christmas, therefore, on 25 December, the first mammals appear. The unfolding of the Alps takes place on December 29 in the evening. On the night of December 31 appears the modern man and 5 seconds before midnight lives Jesus Christ.

7.14 The Future of the Universe

Our home galaxy, the Milky Way will not exist forever. In 2.5 million light years away lies the danger: the Andromeda galaxy. This is racing with 400000 kilometres per hour directly to the Milky Way. Will there be a direct collision when both galaxies pass each other? For many years, measurements of the Hubble Space Telescope have been evaluated in order to clarify this issue. The result: Both galaxies race on a single-track railway line to another. The collision is inevitable. This collision takes place in about two billion years. From the ruins then a new giant galaxy is created.

Currently, the Andromeda galaxy appears as a small spot in the firmament. This spot will increase in the next few million and billion years until it will cover a large part of the sky. What happens after that, we can simulate with computers: Because of the great distances of the stars it will hardly give star collisions. However, huge gas clouds arise and from the resulting higher density of matter new stars will be born.

What happens next? We know that the universe is expanding. Will this expansion stop or maybe even convert it to a subsequent implosion? The mathematical formulas of the general theory of relativity Einstein's left to a decreasing expansion or even a subsequent implosion.

The big surprise occurred in 1998 when two research groups published their knowledge-based metrics: An unknown energy (dark energy) makes the escape velocity of galaxies bigger, the expansion is accelerating (see Chapter 7.11). If this accelerated expansion continues, the universe is in the distant future losing in the infinity of space.

8. IS THE UNIVERSE INFINITE?

> *Two things are infinite, the Universe and human Stupidity, but in the Universe I'm still not quite sure.*
>
> *Albert Einstein*

It's the ancient question: Is the universe finite or infinite in extension? The answer is so far open. Would we know the geometric structure of the universe in more detail, we could answer the question.

8.1 Curvature and Geometry

As stated in Chapter 4, there is besides the Euclidean geometry that corresponds to our school geometry, geometries with curved spaces. An example would be the spherical surface.

The cosmos could be curved like a sphere. The cosmologists calculate a global curvature value which could give light on the nature of the curvature. There are three options: The curvature value could be positive, negative or zero. In case of a positive curvature, the universe would be finite, because a positive curvature corresponds to the curvature of the spherical surface and which is finite in extent. If the curvature is negative, we would have a geometry which corresponds to a saddle surface and the universe would probably be infinitely extended. Finally, if the curvature is zero, we have a flat universe. This is equivalent to a Euclidean geometry, as it has, for example, the plane. The plane is infinitely extended, so perhaps even the universe.

Which geometry really exists, cosmologists try to find out by measurements with space telescopes. These are satellites that orbit the Earth and make detailed measurements of radiation (background radiation). So the Teleskopes COBE (1989) and WMAP (2001) were launched into space. From the measured values, you can infer the geometry.

The analysis of the data shows that in all probability the curvature is zero, so there is a flat universe. With a low probability, the curvature could be positive, which corresponds to a spherical geometry.

Fig.15 A torus. The universe could be a torus

8.2 Is the universe a torus?

Most cosmologists believe that the universe is flat like the plane is flat. This corresponds to the Euclidean geometry. This would make it infinitely expanded as the plane or the space. However there could be another solution. There are spatial structures which, although having the geometry of the plane, are nevertheless not infinitely large. Would the universe have one of these structures, it would be Euclidean, but finally extended.

One of these structures is the torus. A three-dimensional torus has the shape of a car tire. (see Fig. 15).

The surface of a torus as it shows the Fig.15, is obviously two-dimensional and smooth, so it has the geometry of the plane. The torus itself is three-dimensional.

How can the universe surrounding us be a torus? For this we need to go one dimension higher and consider a four-dimensional torus. Unfortunately, our view is limited to our three dimensions, since we live in a three dimensional space. A four-dimensional torus, we can not imagine unfortunately. For mathematicians, however, it is completely without problems, to describe such a torus exactly. And so, like the surface of the torus of Fig.15 is two-dimensional, the "surface" of a four-dimensional torus is three-dimensional. We therefore have no plane, but a space in front of us and this space means the volume in which we live.

What would happen if we go on such a torus toward the small size straight? We would walk around the "tube" and come back to the starting point. Even the light circles the torus and we could see us from behind to a certain extent. So we fly with a super-super rocket in a certain direction, we could come back to the starting point after a (probably very long) time. Overall, the universe would not be infinite.

8.3 The Universe: finite or infinite?

Most cosmologists infer from the measured satellite data and believe that the universe has a Euclidean geometry, so smooth like the infinitely extended plane. Thus the universe could be extended indefinitely. However, the universe could also be finite. As shown, the universe could be constructed as a torus. This would also smooth, but finally extended.

Is there a way to find out whether finite or infinite, whether a plane or a torus is present? An indicator of this alternative is radiation, which is known as background radiation and fills the entire space. The detailed analysis of this radiation could provide an answer. Therefore, we try to measure this radiation with ever more precise space-based telescopes.

In 1989, the COBE satellite was launched and 2001, the WMAP satellite with accurate measurement methods. In 2009 "Planck" was launched into space, the readings will be even more accurate.

The background radiation originated about 380.000 years after the Big Bang, when the universe was created in a kind of explosion. At that time the free electrons and the free nuclei came together and were combined to form atoms, such as hydrogen and helium. Was that universe previously opaque due to the plasma, consisting of electrons, nuclei, photons and other particles, it was now clear, after the plasma had condensed into atoms and the radiation had a clear path.

It crosses the room and since we can measure the corresponding radiation. Cosmologists receive radiation with telescopes mainly as microwaves. This radiation has a temperature and has cooled down strongly due to the expansion of the universe.

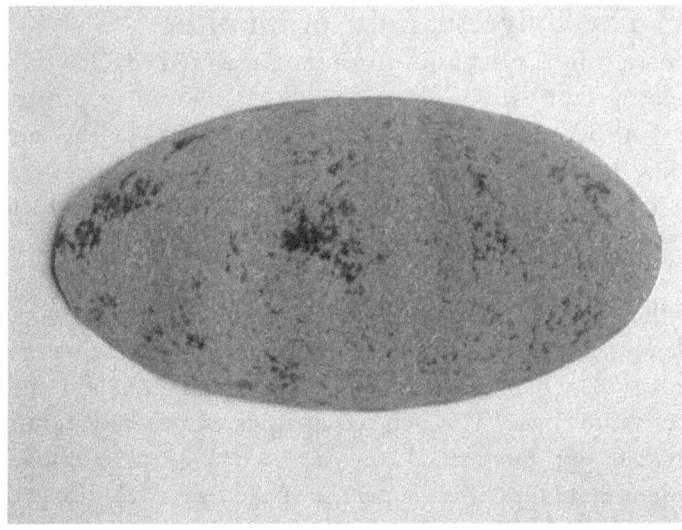

Fig.16: The temperature fluctuations of the background radiation received from the WMAP satellite. (yellow = warm, blue = cold)

The satellite images show a certain local temperature variation, as shown in Fig. 16,. The yellow spots are available for a higher temperature. From this distribution, the structure of the universe can be computed. The computer calculates a direct map of the temperature distribution. If it matches well with the images of the satellite, it was calculated with the correct model. The results of these calculations show that most accurate are the simulations when the calculation was adopted by a space with Euclidean geometry. Therefore, we must assume that the universe is (almost) as the Euclidean plane or not curved space. Thus, the space should be either infinitely extended or be like a torus finit.

The German researchers Frank Steiner and his team

simulated the torus universe on the computer. The results agree very well with the measurement results of the satellite, in some cases even better than the simulation data of an infinitely extended universe. However, the results are not yet sufficient to make the final decision if the universe is finite or infinite. Future accurate measurement data are sought. The German cosmologist Steiner said: The observations will decide it,

9. HOLES IN SPACE?

The universe is a circle whose center is everywhere, whose circumference is nowhere.

Blaise Pascal

The equations of the general relativity describe not only the curvature of space, they allow also strange objects: black holes and wormholes. While we know that black holes exist in space, of the existence of wormholes so far nothing is known. If they exist, we could travel quickly and elegantly into other areas of the universe, in which prevail different times. Travel in the past and the future would be possible.

9.1 Black Holes in Space

Imagine a ball hovering in the space and sucks all the objects in their environment: paper plates, cups, vases, etc.. All these things are swallowed up by this mysterious ball. The writer Gustav Meyrink described such a scene one hundred years ago with impressive artistry. He could not have foreseen that these mysterious objects were discovered a few decades later: black holes.

What are black holes? Let's throw a ball into the air, it comes back as a result of gravity. We shoot with a rifle upward, the bullet reached far greater heights, but it comes also back to earth. We enlarge the launch speed, the achievable heights are greater. Could we shoot with a launch speed of over 11 kilometers per second up, we would wait in vain for the return of the bullet. It would leave the gravitational field of the earth and disappear in space. If we perform this

Where does the space end?

experiment on the moon, a shooting speed of 2.3 kilometers per second would be enough. The speed that is necessary for a body to leave a planet or a for celestial body, is called the escape velocity.

Imagine, we could enlarge the mass density of the Earth. Then the attraction is greater and the escape velocity increases. In a thought experiment we could compress the earth into a small ball. The mass density would increase and the escape velocity also.

If we compress the earth so that it becomes a mini-earth with a diameter of 2,000 kilometers, it would have an escape speed of 28 kilometres per second.

With a diameter of 20 kilometers, there were already 9000 kilometers per second and at a diameter of 20 centimeters about 90000 kilometers per second. With a diameter of about 17.8 millimeters (ie an earth radius of 8.9 millimeters) the escape velocity would be the speed of light, ie 300,000 kilometers per second.

Would it be possible to further continue the shrinking process, the escape velocity would exceed the speed of light. The light could no longer leave the mini-mini-earth. From the outside, the earth would not be visible. We have a black hole.

Black holes are so characterized that their mass density is so great that light can no longer escape. Moreover, their gravity is so strong that they attract all matter around and the attracted matter disappears, never to return into the black hole.

The first person who thought about what would happen if the mass density of a star is so great that the escape velocity is greater than the speed of light, was the British John Michell 200 years ago. He found out that, if the radius of the sun would be 500 times larger, light could no longer escape. In 1784 he reported it in the philosophical treatises of the Royal Society.

Similar thoughts went to the Indian student Subrahmanyan Chandrasekhar. He wanted to study at the famous astronomer and physicist Arthur Eddington in England and took the long boat trip from India to England to think about what the consequences would be if you'd shrink a star or planet. The density would be greater and hence the escape velocity. Light could not escape, the star looks from the outside black.

It originated the term "black star" or "frozen star". The term "black hole" was created in 1967, when John Archibald Wheeler reported a lecture about these objects. He called them "gravitationally completely collapsed objects". An unknown listener suggested the name "black hole" in front and the name stands since then for these objects.

How far you have to compress the earth, the moon or the sun, so that they become black holes? The astrophysicist Karl Schwarzschild, a professor in Göttingen (Germany), and later director of the Astrophysical Observatory in Potsdam was looking for a formula to calculate. He found it 1916 shortly before his death, which overtook him as a result of war suffering. The radius to which you have to squeeze a star or planet, was named in his honor "Schwarzschild radius". For our earth it is 8.9 millimeters and for the sun about three kilometers.

Black holes exist in space, for example in the center of our galaxy with 4 million times more mass than the sun. Black holes are the burnt residue of earlier stars. In our neighbouring galaxy, the Andromeda nebula, there is a black hole with a mass that is 50 million times greater than the mass of the sun.

Near a black hole, the gravitation takes unimaginably large values, therefore all matter within a neighbourhood is sucked mercilessly. Would an astronaut approach to a black hole, his clock would go slower and slower, because - as described - gravity slows time. The space would its curvature always increase. However, the astronaut would not perceive time reduction. In his perception, the time runs completely normal. But if we could watch the flight of the astronaut from Earth, we would find that with him everything is in slow motion. When he reached the edge of the black hole, the time - observed from us - stops going. The astronaut would not live to see that, the high gravity would tear him and his spaceship before.

9.2 Wormholes

Many of us know it: astronauts of science fiction scene, depicted in film, television and literature. They fly through the space and meet now and then to aliens who live on other planets. Curiously, these residents look so similar like we, perhaps with small variations such as yellow hair or washboard giant ears. The most amazing thing is that our astronauts are able to converse easily with the aliens, without interpreters. Now and then it also comes to fight, but our imagination astronauts win always.

How came our astronauts into such distant worlds. Would they, for example, fly to a planet that belongs to a star nearest to us, they needed by conventional reputable calculation thousands of years. Our astronauts but do it by flying like a taxi in a high electronic capsule with huge switchboards from planet to planet. The solution is simple: they use wormholes.

Wormholes are direct connections between different areas of

the galaxy that you like by flying through a tunnel. Would you climb a wormhole near Earth, you could theoretically come out at the other end of the galaxy. A route that is thousands of light-years long they overcome in a few days.

The equations of general relativity allow these tunnels. Does it follow that they really exist?

What is a wormhole? This results in a space, which is curved. Since the universe exhibits a curvature, wormholes are - at least theoretically - possible. Since we can not imagine a three-dimensional curved, we consider the problem a dimension deeper, so in a two-dimensional curved space, such as the spherical surface.

For simplicity we imagine an apple. Its surface is two-dimensional and curved. Will a worm from one point of the apple surface crawl to another, he has a (for a worm possibly) long road to travel. Faster it comes to goal when he uses a tunnel through the apple, and it follows from point to point along a straight line and thus it has a shorter distance. We have a wormhole.

We apply this model to three dimensions: The universe is curved like the apple surface. So you can - at least theoretically - leave the room through a tunnel and this tunnel opens into another area of the galaxy, where optionally also prevails a different time.

A wormhole is thus a tunnel in the universe through which you can leave the room and come out again elsewhere. There are not only the spatial coordinates are different, but you ended up in a different era, so in the past or in the future.

Are there wormholes in space? The theory of relativity allows for it, but it is highly doubtful whether they actually exist. Not everything mathematics pretending exists real. Presumably wormholes are an interesting gimmick with the equations of general relativity. Kip Thorne, dealing intensively with this matter, says: "wormholes and time machines are now rejected by most physicists."

10. THE ANTHROPIC SPACE

We absolutely must have room for doubt,, otherwise there is no progress

Richard P. Feynman, quantum physicist

Had the universe only to be created to bring forth life, it would seem oversized in its unimaginable extent. On the other hand, when life is created, the expanding of the universe at the given physical laws for billion years makes its gigantic size forcibly.

The physical laws that govern the dynamic history of the universe, are based on parameters and fundamental constants. If you would change these constants only slightly, a universe would almost always come into being without life. The parameters and fundamental constants are set so that inevitably lives must arise. Cosmologists refer to the "anthropic principle"

10.1 Is the universe oversized?

The universe contains in its gigantic size 1 billion times 1 billion stars and among these many planets. Should all this be created to bring forth life on a planet called Earth? If so, it seems to be oversized grotesquely.

Suppose a creator wants to create life on a planet like Earth, using the known physical laws. In a big bang initially elementary particles are created that form into atoms. The first atoms are light atoms such as hydrogen and helium. Heavier atoms are formed in sun-like stars by thermonuclear processes, such as carbon, oxygen, phosphorus. Some of these stars explode as supernovae and hurl heavy atoms into space. Many of these atoms are part of life.

The physical laws are structured so that the flow of all these processes must take many billions of years. During this long period, the universe must be expanding and renewing, because otherwise the gravity would have pulled together the newly formed atoms and the universe would collapse even before life could have arisen. But if the Universe expands many billions of years at a high speed, it must assume necessarily a gigantic size. In the vast rooms are created atoms, from these stars and then billions of celestial bodies.

After heavy elements were created by supernova explosions, it took again many billions of years before would have been formed from these elements and planets on which life could arise. Even now, the universe had to expand, so as not to coincide. This expansion continued to the present size of the universe.

The fact that we exist, therefore presupposes the size of the cosmos, the billions of galaxies, the vast periods of time. This gigantic size of the universe is prerequisite so that life can arise. The universe is therefore not oversized, but it has just the right size and dimensions, so that life could arise on our planet and perhaps on other planets. If life should arise for the universe quite different physicochemical laws would have been necessary.

The Paris-based Australian theoretical physicist Brandon Carter has given in his treatise "The Constants of Nature" an estimate of the amount of time that is at least necessary that under the Evolution can occur higher quality of life. This time coincides with the age of the universe.

10.2 The enormous coincidences in space

Since its creation some 14 billion years ago the universe was expanding ceaselessly. Like a balloon that is inflated, it expands in every second. The galaxies are driven away from each other by a mysterious energy. Since the nineties of the last century, we know that the expansion of the universe runs faster.

Antagonistic to this there acts another force acting in the opposite direction which likes to merge the galaxies: the gravitation. Both forces are necessary and keep the universe stable. If there were only the gravity, the universe could not develop. If there were only the expansion, all matter particles would escape into the vastness of space and life would never have arisen. Expansion and gravity keep the universe in balance.

How have expansion and gravitation be designed in their strength, so that the scales are in balance? Too much expansion would push the universe apart so quickly that life never could have arisen. By too much gravity the universe could as early collapse, that the time was not sufficient to create life.

The strength of gravity depends on the density of matter. The more matter, the stronger the gravity. It can therefore also be formulated as: How large must be the density of matter after the Big Bang, so that the universe is expanding and so that later life is born?

The cosmologists can calculate the mean density of matter that must prevail one second after the Big Bang, so that the universe can produce life at a later date. A deviation from this density to only

Where does the space end?

000 000 001 0,000 000 grams per cubic centimetre

had created a space, that would be expanded either too fast or too soon would collapse. So it was an accurate fine-tuning necessary if life is to be built billion years later.

It is similar to a pendulum clock that you set at the beginning of the year so that the correct time is always displayed. The pendulum must be exactly 99.396082 ... centimetres long. Is it even a tenth of a millimetre too long or too short, the clock would be at the end of the year gain or lose 10 minutes. So it's a huge coincidence that exactly the right density of matter arose.

The coincidences in the development of the universe are varied. For example, the ratio of gravity and electric power. Gravity is the 10^{-36} times the electrical power. Suppose that we could play with the parameters of the universe and in a simulation enlarge the gravity so that it is the 10^{-26} times. With the aid of cosmology we can calculate exactly which properties such a universe would have. All the stars would be much smaller than they are. Approximately 10 million would put together a mass corresponding to that of the moon. Everything would proceed much faster. After one year, the stars would be burned and life could never develop in such a universe.

Another "coincidence" is the life of the neutrons. An atomic nucleus consists of protons and neutrons, the neutrons are vital for the atom. Without neutrons atoms would be unstable.

Immediately after the formation of the universe in the Big Bang, the temperature in space was so high that no atoms could be formed. The protons and neutrons were flying as free particles through space. However, neutrons decay already after 11 minutes. After 11 minutes, therefore only protons were left and stable atoms would never have been arisen and it would not have given us. The only way to save the neutrons was that the temperature in the early universe would fall so fast that because of the lower temperature neutrons and protons could come together within 11 minutes.

In the moment when a neutron is captured in the nucleus, it is saved. Its disintegration no longer takes place. It is therefore a further stroke of luck that the temperature of the universe was so strong reduced that atoms with neutrons in the nucleus could arise in time.

Imagine a universe where we could design each natural constant and mass fraction like to set screws. We change one of the variables a bit only. The so altered universe may bring perhaps forth no life. In that case, we correct, by changing other parameters in order to "save" the universe. If we continue like this, we will probably end up in a maze and eventually nothing works.

All physical constants are apparently coordinated. If you change one, you may create a universe in that life is a stillbirth with respect to life. It's like a melody: If you change a single note, you destroy the harmony.

That in the universe was created life, was possible because the fundamental constants had the correct values. Humans fit to the universe and the universe fits to humans. This is the statement of the anthropic principle, first was described by Brandon Carter in 1973.

Cosmologists use the anthropic principle by - because life exists - they close back how constants of nature or mass fractions in the early universe must have been set.

10.3 Universe: accidental or creation?

In May 2007, a museum was opened Petersburg in Cincinnati, USA, which should prove that the world was created in 6 days. Adam and Eve live there peacefully with dinosaurs together, although the dinosaurs were extinct in the origin of man.

On the other hand assert fundamentalist evolutionists that the origin of life based solely on chance. Everything arose mechanistically, a creator is superfluous. As an example of atheists Richard Dawkins may be mentioned with his book "The God Delusion".

Regrettably speak both sides - creationists and fundamental evolutionists - from the non-science of the other side, although neither side can prove their position exactly.

That life originated on evolution, is undisputed. However, there are in the evolutionary course things that are not rationally explicable.

Thus, it is unclear from where the first divisible biological cell came. There are also complexity jumps that are not explainable solely by the evolutionary model. Such leaps in evolution are, inter alia, described by Professor Michael Behe of the University of Pensylvania. The mathematician William Dembski calculated the probability of purely random evolutionary development of highly complex DNA strands, the carriers of life. Its result as a development only by chance is impossible.

So the alternative "Creation in six days" and "Evolution" is no real alternative. Both alone can not explain the origin of space and life. The theory of evolution is correct, but it does not explain everything. About mythic representations such as the creation story writes the quantum physicist Werner Heisenberg in his book "Der Teil und das Ganze" (The part and the whole): "The language of religion is with the language of poetry more closely related than with the language of science. .... But that does not mean that it is not a genuine reality. " (see. [H96])

11. SPOOKINESS IN SPACE

> *The more successes the Quantum Theory has, the sillier it looks*
>
> *Letter Albert Einstein to Heinrich Zangger on May 20, 1912*

Some time ago, a bomb from the Second World War was found in Munich in Germany. It seemed impossible to defuse the bomb conventionally, therefore it should be controlled to explode. As expected, buildings were damaged in the neighborhood.

Imagine the explosion in Munich would have caused damage in a street in Hamburg, some hundred kilometres apart. That would be pure spook and unexplainable.

Since a few decades we know that this spook is given in the microcosm. The physicists call it non-locality.

When the first suspicions arose that there might be such a thing, Albert Einstein called it "totally unacceptable", called this idea "spooky" and didn't believed in it. Since the eighties of the last century we know from experiments that this in the microcosm really exists.

11.1 The location of the room

This "Spooky" is found in the non-locality of the room. One

of the important properties of the room is its locality. If I throw a stone into a lake, then he is making waves near the point of impact. These waves are spreading. The stone affects the lake just nearby, so locally. Every action within the space has consequences only in the neighborhood, and propagates outward. Here, the propagation speed is always smaller or equal to the speed of light. The location of the room was therefore a natural property that resulted from the prevailing laws of physics. Every action in the area has an impact on its immediate local environment.

There was a lot of excitement when some quantum physicists claimed that in the microcosm the locality does not apply. That would be as if someone threw a stone at Oxford in the River Thames and its waves arise at the time of impact in the Thames at London. Non-locality was supposed to prevail in the microcosm. Albert Einstein refused to accept the non-locality. Later experiments showed that the non-locality in quantum physics actually exists.

11.2 Entangled photons

The non-locality refers to elementary particles such as photons, which were eventually combined in an atom. Physicists speak of entangled photons. They can be created, for example, by producing a photon in a crystal, which disintegrates into two photons and which fly away in different directions. Both photons are not independent, they behave like twins. The one photon "knows" to the other, although both are light-years apart.

This will be explained in more detail: photons or light particles have a polarization which is an excellent level of vibration. Physical conservation laws require that entangled photon polarizations of the two photons are perpendicular.

The direction of polarization of a photon can be measured. Quantum physics shows that the measured value of an elementary particle is set only in the measurement, but before it is completely undetermined, so absent. Only measuring takes out a value by chance from the possible measured values and assigns it to the particle. One could say that only when measuring then particle "decides", which measured value is to accept.

This discovery was one of the revolutionary discoveries of quantum physics, which was initially not accepted by many physicists. Measuring is watching and the quantum physicist Werner Heisenberg wrote in 1927 in the Journal of Physics: "The track is produced only in this way that we watch it" He meant here the trajectory of an electron.

Suppose now, two entangled photons move at the speed of light away from each other and they are already thousands of light years away from each other. One of two hits on earth, and we measure its polarization. The second distant photon is entangled with the measured particles and quantum physics demands that its polarization is on the polarization direction of the first photon vertically. The second photon "knows" that his twin is measured and sets at the same moment his own polarization so that physics remains accurate.

That sounds so abstruse that many physicists were not willing to accept these ideas, including Albert Einstein. This also contradicts the special theory of relativity, according to which the speed of light is the maximum speed.

Einstein kept trying to show mental constructs that should disprove the non-locality in the photon. He then wrote: "The conclusion ...... can only be avoided by assuming either that the measurement of S1 (first particles) the real state of S2 (second particles) changed telepathically... This seems to me quite unacceptable. He described the phenomenon as "spooky ".

In 1982 succeeded the French physicist Alain Aspect and his team in Paris to demonstrate the non-locality in quantum physics. Later experiments, carried out by other research teams, confirmed this. The latter could even show that there are also entangled atoms with similar characteristics.

For the non-locality in space and other statements of quantum physics there are no clear explanations and so could quantum physicist Richard P. Feynman, who in 1965 received the Nobel Prize, formulate the sentence: "It is safe to say nobody understands quantum physics". Feynman loved to declare a wide audience the oddities of quantum physics with humour. From these presentations out his book was titled: "Surely you're Joking, Mr. Feynman" .

11.3 The separability of space
Two entangled photons behave simultaneously while measuring. In other words, if I measure here, arises at the same moment light-years away in the second photon the associated measurement value. If two things that are far away from each simultaneously respond, they behave as a unit. They seem to be connected in any way.

If I measure a first photon the measurement information arrives in the second photon at the same moment. One could formulate that the information "measured value" propagates to the second photon in an infinite speed.. This was the reason why Albert Einstein and other physicists did not recognize the entanglement of particles, because this idea clearly contradicts the special theory of relativity, according to which there is no speed that is greater than the speed of light.

To illustrate the phenomenon, we look at an iron bar. If I move the bar at one end, the displacement at the same moment at the other end is done (This is true only in principle, but that we want to neglect). The reason: The bar is a unit that responds as a whole. The information "Move" propagates with infinite velocity from one end of the rod to the other. If the signal would propagate with only a finite speed, you could temporarily stop it on the way, such as when it has reached the centre of the bar. Now one half of the rod is completely unaffected by the signal, while the first half has already received the signal. So we could divide the bar into a first part and a remaining part. We have thus the space that occupies the rod, divided or separated. Physicists speak of the separability of the space.

That space is separable resulted from the special theory of relativity. The separability was solid until the discovery of the laws of quantum mechanics. It says that every area of the room is divisible. The behaviour of entangled photons shows that the separability in the microcosm apparently does not apply.

12. MATHEMATICAL SPACES

Do not worry about your difficulties with the mathematics; I can you assure you that my difficulties are even bigger

Albert Einstein on January 7, 1943 to the student Barbara Wilson

No science is engaged in such detail with the concept of space as mathematics. The central discipline within mathematics, which deals with the laws of space, is the topology. This is the doctrine of the topos, this means the Greek word topos for space, region, area. There are one-dimensional spaces, two and three dimensional spaces, etc. Since the chaos theory, we know that there are even broken-dimensional spaces such as a 1.6-dimensional or 3,7-dimensional space.

12.1. Linear spaces

For mathematicians, the notion of space is one of the most important tools in the description of the predictable world. Amounts can reveal structures that resemble the structures of spaces. These amounts may be: sets of numbers, sets of functions, complex numbers, set of all numbers divisible by seven, etc. Their space-like structure is referred to a "linear space". It has been recognized that the basic structure of such an amount has the structure of a linear space, you can apply all the laws that are inherent to the spaces on these quantities and gain additional insights.

To find these laws of space, we first consider the plain. This is a two-dimensional space. From school is known that one can describe and represent this area in a coordinate system.

Each point in this area (see Fig. 17) is characterized by two numbers: the x-coordinate (abscissa) and the y-coordinate (ordinate). So are, for example, on the drawn line in Fig.17 are the points $x = 2$, $y = 1$ or $x = 4$, $y = 2$, how easily seen. We write shortened $A = (2,1)$ and $B = (4,2)$.

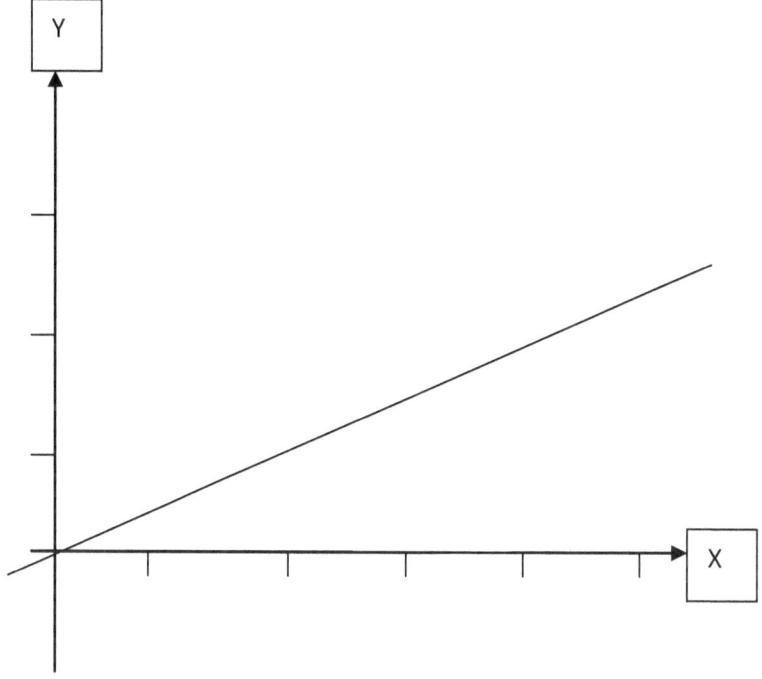

Fig. 17: Coordinate system with abscissa and ordinate

Points of the coordinate system can be multiplied by a

number and can be added together. Examples for the point A = (2,1) and B = (4,2) :

$$6 * A = 6 * (2.1) = (12.6)$$

$$5 * B = 5 * (4.2) = (20.10)$$

$$A + B = (2,1) + (4,2) = (6,3)$$

It is thus multiplied and added by coordinates. Of course, the subtraction is:

$$A - B = (2,1) - (4,2) = (-2, -1)$$

All points that arise in these calculations, are back in the plane, in the coordinate system.

That you can multiply the points of a plane with a number or add two points, is also a characteristic of the three-dimensional space, of the one-dimensional space (the line) and the property of arbitrarily dimensional spaces. Mathematicians refer to these areas as "linear spaces". You define a linear space as follows:

Definition 1
A set R with the elements (points) A, B, C, etc is a linear space if the following applies:

1. If the points A and B are in the set R, then also A + B.

2. If A is a point in the set R, then A * λ, where λ
   is an arbitrary number.

Regarding the multiplications usual rules like A + B = B + A are guilty., what is not to be discussed here. It is easy to check that the above definition for our two-dimensional space is given.

12.2 High-dimensional spaces

Use the above definition of spaces you can now very easily three-, four- or even 40-dimensional spaces calculate. Although our imagination extends only up to the three-dimensional spaces, physicists regard the four-dimensional space, the fourth dimension is time. The representation of arbitrary dimensional spaces is no longer a problem.

Let's start with three-dimensional spaces. We now need a three-dimensional coordinate system. The three dimensions are represented by three axes with the coordinates X, Y and Z. These axes are perpendicular (orthogonal) and are shown in Fig. 18,. The x-axis is perpendicular to the plane of the paper and thus is represented in perspective.

Each point in this space has the three coordinates (x, y, z). If the point (3,2,4) is to be inscribed, you go 3 units in x-direction, 2 units in y-direction and 4 units in the z direction. For example, in Fig.18 it is easily seen that the definition 1 in Chapter 1 is met, therefore we have a linear space. Each point (x, y, z) belongs to the space. Mathematicians refer to this point as a vector (in a slightly modified way).

If the description of three-dimensional spaces is so simple, then also for four-dimensional, then for eight-dimensional or even for hundred-dimensional spaces.

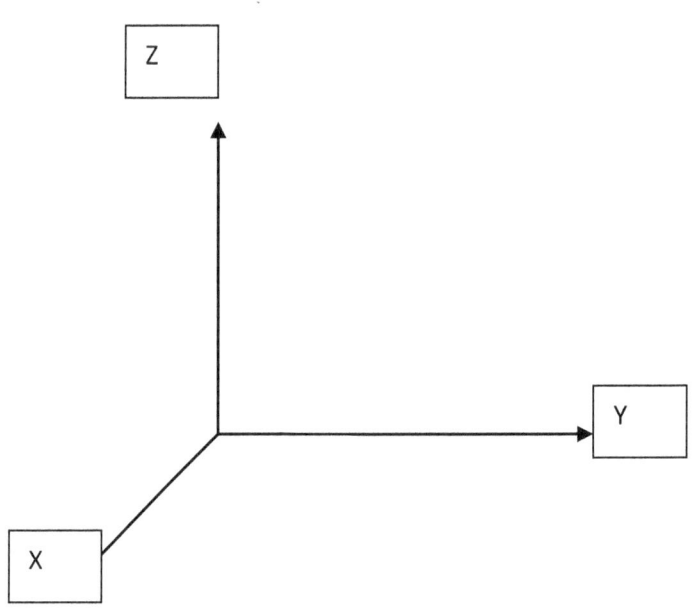

Fig.18: Three-dimensional coordinate system. The x-axis is drawn in perspective and is perpendicular to the plane of the drawing.

In a four-dimensional space we write the points as (x, y, z, w), that is, for example, (3, 6, 2, 3). For this you need a four-dimensional coordinate system. Nature has created us only in a three dimensional space and therefore also our imagination is only in three dimensions. A four-dimensional coordinate system exists, but we can not imagine it. But mathematically it is easy to work with four-dimensional spaces.

After this abstraction on four spatial dimensions, it is easy to regard even eight or 100-dimensional spaces. A point in the eight-dimensional space would be, for example,

Where does the space end?

$(1,5,3,-3,5,9,7,8)$.

For the solution of systems of equations with thousands of unknowns mathematician work with points (vectors) in a coordinate system with thousands of dimensions. For mathematicians, it is no problem, even to work with infinite dimensional spaces.

12.3 Amounts as linear spaces

Space structure have not only spaces in the above sense but also amounts, in which the layman would not expect it, which has nothing to do with "spaces" at first glance. If you have recognized a space-structure, you can all the laws that relate to mathematical spaces apply to these amounts and you gain valuable insights.

For example, the set of all numbers divisible by 5 has space structure. Because: Multiplying a number divisible by 5 with a completely arbitrary number, the product is again divisible by 5. If you add two numbers divisible by 5, the sum is again divisible by 5. So there are the laws of Definition 1 in Chapter 1. Thus, the amount constitutes a linear space.

The set of all (real) numbers forms a linear space. Taking only the even integers, then they form a linear space, because Definition 1 applies. The amount of odd numbers like 1,3,5,7 etc however do not form a linear space. Also the set of prime numbers is not a linear space, since the sum of two primes is not always prime.

An important set for mathematicians is the amount of (defined in an interval) functions. We know functions from school mathematics or from calculators. Example: $y = x^2$, $y = \sin(x)$, $y = \log(x)$, etc. Adding two such given functions arises a new function. The multiplication by any number results in a new function, ie Definition 1 is satisfied, the functions form a linear space, they have a spatial structure.

12.4 Fractional dimensions

With the spaces previously considered it concerns to n-dimensional spaces. Here n is always a positive integer. If n is a decimal number, such as n = 5.7 or n = 1,267, one can imagine a corresponding space. Mathematicians found spaces with such dimensions. They speak of spaces with a broken dimension.

Imagine a silk thread. It is like a curve one-dimensional. Is it possible to lay it out so that it covers all the points of an area, for example, a piece of paper? In this case, the structure thus created like a piece of cloth would be two-dimensional. That this is actually possible discovered in 1890 the Italian Guiseppe Peano. This curve is known as the "Peano curve".

Later there were found other curves that fill like the Peano curve a whole area. These curves were preparing mathematicians some headaches, because they are one-dimensional curves but they fill an area, which is two-dimensional.. The mathematician H. Poincare found for these structures the term "gallery of the monsters"

It was further complicate, when the curve indeed fills an area, but not all points of the surface are detected. An example of this is illustrated in Figure 6 Koch curve (see. Chapt..4.4).

Using mathematical methods, which are not to be discussed here, one can determine the dimension of the Koch curve if n goes to infinity. One experiences a surprise: This curve has the dimension of 1.26285, ie a fractional dimension. The

Sierpinski triangle of Figure 7 has the dimension $\log_2 3 = 1.58496$. Benoit Mandelbrot found in 1977 more structures with broken dimensions. The dimension of the coastline of the UK was given as 1.26.

13. A BRIEF HISTORY OF SPACE

Time is movement in space

Joseph Joubert

Initially the room was conceived as a kind of container that provides the boundary condition for the content, later there developed a relational understanding of space, in which then space is relative to bodies. Finally, the time has been integrated as a fourth dimension to the concept of space by Einstein, there was the space-time continuum.

13.1 The space in antiquity

The Greek philosopher Aristotle (384-322. BC.) believed in a finite space. Outside this space there can be nor enough time neither space for his opinion. All movements of the heavenly bodies are caused by a first unmoved mover. The circular motion of the stars he interpreted as a quest for eternity and continuity.

Later there arose the Euclidean geometry. Euclid lived from about 360 BC. to 280 BC., he is probably the one that summarized the knowledge of his time in terms of the geometry. He wrote in 13 textbooks, the "elements". The elements are the most successful textbooks ever and were used in England even in the 19th century as textbooks.

Titus Lucretius Carus (about 97-55 BC) believed in an infinitely extended space. He explained it by the fact that people, if the universe were finite, could go to an end and there could hurl a lance. But since there is no reason why the lance should stop its movement when it is thrown in the direction of the space limit, it can not give this limit, therefore the universe is infinite.

Similarly, Pope Simplicius expressed (5th century). Would that all finite, you might go in his opinion to the border and reach out your hand. If you feel resistance, it would be a limiting wall. If there is no wall, you stretch out your hand again. Since one can endlessly continue, the universe is infinite.

13.2 The heliocentric and geocentric worldview

The ancients concluded from the movement of the sun in the firmament, that the sun moves around the Earth. At this called geocentric model view also believed the Greek mathematician and astronomer Aristarchus of Samos (about 310-230 BC, more precisely. Aristarchos). Later, however, he changed his mind and declared that, conversely, the earth moves on a circular path around the sun. (heliocentrism). He also noted that the sun would be far larger than the Earth. Unfortunately there are no writings, which date back to Aristarchus and so got his idea of the heliocentric worldview into oblivion. 1800 years later the heliocentric wordview was rediscovered by Nicolas Copernicus again.

Claudius Ptolemy was a librarian at the famous Library of Alexandria and lived from 100 to about 175. He described a geocentric model, in which the sun revolves around the earth. His orbit calculations were amazingly accurate, so that his worldview was able to maintain for centuries until finally penetrated the Copernican worldview. About a thousand years the Ptolemaic worldview was common knowledge. His handbook on astronomy was instrumental until to the 15th century.

In the 15th century the priest Nicholas of Cusa (1401-1464) surmised, that not the earth but the sun stands still and the earth revolves around the sun. However, he could not develop his theory mathematically.

1543 finally, it was the chancellor of the cathedral chapter of the diocese of Warmia in Frauenburg Nicolaus Copernicus, who described - possibly inspired by Aristarchus - in his treatise "De revolutiobnibus Orbium Coelestium" a heliocentric worldview. Copernicus lived from 1473 to 1543. His theory was based on observations of planetary motion and resulted from considerations relating to a calendar reform. All the planets revolve around the sun and the earth rotates around its own axis.

Copernicus was not fully convinced of his model, whose accuracy he could not prove. Moreover, he feared a disgrace and kept his record for 30 years in the drawer. In the year of his death his work appeared. His theory was taken up by his contemporaries less than heresy, but more than fantasy. It appeared counterarguments as: if the earth were hurtling through space, you would have to feel a headwind - or - objects that fall on the earth would have a sloping trajectory. Luther should - according to student statements - have said: "The fool wants to turn me the Astronomia."

A more serious counterargument to Copernicus model was the lack of parallax. The point is that when circling the Earth in winter one would see the fixed stars at a different angle than in the summer. This change in angle was not discernible. So the earth would have to stand still and not the sun. The fact that the fixed stars are so far away that a parallax is not measurable, was unknown. Only in 1838 Friedrich Wilhelm Bessel was able to measure the parallax of a star

Where does the space end?

An advance in the exploration of space was in 1608, the invention of the telescope by the Dutch spectacle maker Hans Lipperhey. This Dutch telescope instrument was used by the Italian mathematician Galileo Galilei (1564-1641) and he discovered the moons of the Jupiter and the phases of Venus. Galilei entered openly the Copernican heliocentric world view and was not entirely innocent by his excessive zeal in mind that he was banned in 1614 by the Pope. First Galileo was admonished, but the dispute continued. 1633 it came to the process that ended with the abjuration of Galileo and his condemnation. He was sentenced to indefinite detention, but he was able to spend it in his country house. In addition, weekly penitential psalms he was imposed, but his daughter, a Carmelite sister, took care of this.

At times it was incorrectly assigned that Galilei invented the telescope. A trial observer, the Jesuit Juan Valdez, wrote to a friend about the telescope: "Galilei's invention, the telescope, with which you can still distant enemy troops move into a previously unused near to the observer."

A significant improvement of the Copernican worldview was given by the imperial court astronomer and mathematician Johannes Kepler in Prague (1571-1630). Kepler found three laws which affected the planetary orbits. He replaced the circular orbits of the planets by elliptical orbits. The Kepler's laws are:

1. The path of a planet is an ellipse, whose one focal point is the sun.

2. The radius vector connecting sun and planet sweeps
in equal times equal areas.

3. The squares of the orbital periods of two planets are proportional to the cubes of the major semi-axes

Kepler was deeply religious and looked into the laws found by him the expression of a world harmony, which had placed in the Creator. Kepler believed in a finite universe.

13.3 From Newton to Einstein

Isaac Newton found in his work Naturalis Principiae Pricipia Mathematica a new approach to the laws of mechanics, which up to today hold their validity, though they are not quite exact, as later Albert Einstein showed.

Newton assumed an infinitely extended space and the time in absolute value. It's like a great clock that somewhere in the universe for the entire room pretending an absolute time, which is valid in all points of space regardless of their location (see chapter. 1). In this space, the Euclidean geometry is valid, it has three dimensions and is flat.

Non-flat rooms as the spherical surface that are curved in itself, was the subject of the mathematician Bernhard Riemann (1826-1866). Riemann was one of the most outstanding mathematicians of his time. He taught in Göttingen (Germany) and established different mathematical disciplines, including the "Riemannian geometry". This describes the properties of the non-Euclidean geometry, that are curved spaces. Riemann's merit is to transfer the laws of these geometries to any sized rooms. The spherical surface is two-dimensional and curved. Just so one can describe a curved space in three-dimensions.

Riemann considered curved structures with any number of dimensions and introduced the concept of curvature (the curvature tensor). A special case would be the three-dimensional curved space which is not inconceivable, but mathematically very accurately describable. Riemann died at

39 years in Italy from tuberculosis.

The Riemann geometry was a template for Albert Einstein, when he in 1916 published his General Theory of Relativity. Einstein claimed that matter and energy cause a curvature of space. He set up equations which link the space and time on the

.

$$R_{\mu\nu} - 0.5*R* g_{\mu\nu} - \Lambda* g_{\mu\nu} = 8*\pi*G/c^4 * T_{\mu\nu}$$

Einstein's famous equations that established the General Theory of Relativity

one hand and matter and energy on the other hand. The equations contained geometrical factors, from which you could read the geometry of space. The result: the space is curved

14. SPACE SPECULATIONS

*Nothing prevents the soul so much
to the knowledge of God as
space and time*

Meister Eckhart

Time and space are the coordinates that define our lives. Since space and time are not set absolute and rigid but are shown variable and always changing, one could ask the question whether there is a change, in which space and time dissolve or disappear completely. This may seem illusory and strange sound, but a photon speeds finally without time and space through the universe and black holes exist also without space and time.

It should be noted that the arguments presented in this chapter are possibilities, but not proved statements. In this sense, the statements are speculative. Imagination and speculation may find a place in the field of science entirely if they do not contradict the basic equations and if is expressly pointed out that it is possible, but not proven..

14.1 Omnipresence

We fly with a rocket through space. At high speed, the time goes slower and the space becomes smaller in the direction of flight. Suppose we could incessantly enlarge the speed of the rocket. What happens? Distances in flight-direction get smaller and smaller. Suppose further that we could obtain the greatest speed that is possible, the speed of light. Now the time does not only stand, but the distances in the direction of flight have shrunk to null. We are both at the starting point and the destination of our trip, the flight time is zero. This is the situation of the photons, the particles of light, flying with the speed of light through space

If we could observe a photon on his journey from the Sun to the Earth, we would be able to trace it about eight minutes on its 150 million km long road. This is true when we observe it from Earth. Quite different from the vantage point of the photon: For the photon, there is no time, every time is null.

When it starts it is the same time at the destination, because the Sun-Earth distance is zero for the photon. The dimension in the direction of movement is no longer available. It is at the same time on the sun and the earth. For the photon there are therefore only two spatial dimensions left, namely perpendicular to the flight direction.

How would that look like in a spherical wave, ie a spherically expanding wave, which spreads into all three dimensions? Suppose that there is such a wave. As the wave spreads in all directions with the speed of light, it knows no spatial dimensions. For it there is only one point where it is located. The whole space is shrunk to a point. It is omnipresent or ubique, thus simultaneously everywhere. Of course there is for it no time. If we describe the timelessness as eternity, it is -in its own system - omnipresent and eternal.

People who have had a near-death experience, who have died and were then revived, report sometimes about an experienced omnipresence directly. Werner Barz, a roofer from Bavaria in Germany, reported in YouTube (call under the heading: "NDEs 2015"). As a result of an infection he had died, but was revived at the hospital. About his process of dying, he reported: "I felt like I stretched me beyond the room, beyond the earth, in the universe inside. I do not know how to do that, but it was true. If I spread out my hands I felt at my fingertips that I touched the end of the universe. It was a mental and physical perception. It was all one. I was no longer I, but in the whole area of the light there were sensations. The room was gone and the time was gone. "

14.2 Wholeness and unity

In Chapter 11 we considered entangled photons. These are photons, which were together in the past (for example in an atom) and then flew off in different directions. At the moment, when you measure a property of a photon, arises as if by magic a corresponding value in the second photon. This happens also when both photons are light years away from each other. (see.chapter11.2). It is like if the photons know from each other and react like twins.

Apparently the two photons react as a whole. They could form a unit in any way in any by us not comprehensible manner.

To get a possible understanding, we consider a plane. Two points of the plane may be entangled, that means, they react simultaneously like twins .

So there would be a mysterious connection between them.

A connection we could construct, if we assume a connecting line outside the plane. Something like a compasses, which marks at the level of two points: the compass point and the mark point. Both points are connected by a bridge outside the plane, therefore "entangled". Transferred to the area this means that a possible connection outside the perceptual space runs. This apart spatial imagination would be a possible explanation, but of course purely speculative and scientifically undetectable.

14.3 Temporal entanglement?

Space and time are equal from the point of view of physics. If there is a spatial entanglement, then could there be a temporal intertwining? In this case time-shifted operations should be related, which ultimately means that operations of the future or the past would have influence on the present.

So far, such phenomena were not observed, so that we are in the realm of pure speculation. Nevertheless, it is not without interest to regard the opportunities opened up by such an idea.

The quantum physics leads to the thought that future extends into the present. This was proposed by an interesting idea of the Quantum-Physicist Werner Heisenberg.

Heisenberg participated often at colloquia of the Institute for Behavioral Research near Munich. One of these events was about biological evolution. The following comparison was made:

The development of life on Earth is comparable to the development of tools. As an example, look at the development of boats on a lake. Initially there were only rowing boats. Eventually someone came up with the idea to mount a sail and after a certain time you only saw sailboats. The next step was the invention of steamships, which largely displaced the sailing and rowing boats. Finally, the motorboats came. Does biological evolution occur in this sense?

Heisenberg said that the comparison has a point is inaccurate.

After Darwin Evolution works through constant and continuous change for the better. But you can make to a sailboat so many alterations, it will not change to a steamship. Heisenberg writes: *"I tried to imagine what comes out when the comparison would be taken more seriously than it had meant and what would have to take the place of Darwinian randomness. Could we start with the term "intention"?"*

He concludes that what we generally understand by intention probably is hardly suitable. Rather, one should define the term "intention" from the view of the quantum physics. There, it is believed that in the microcosm the change is described in energy and matter by a wave. Heisenberg writes: *"Since the random change takes place at the level of quantum physics, or at least initiated from here, the wave function is valid and represents what is possible, not what is factual. Could the selection of the possible changes be influenced from the future or from the target be to make the change so that better adjustments to environmental conditions occurs."*

Most present biologists had trouble to adhere to that view. For them - as Heisenberg - atoms and molecules are objects of classical physics, although they recognize the authority of quantum physics. But - as Heisenberg – *"One can occasionally come to very wrong results, if one thinks in terms of classical physics"*.

Appendix

This part of the book is about the creation of Mandelbrot sets (see. Figure 5) with the aid of a computer. These images are based on complex numbers and every hobby programmer can create these images when he has a basic knowledge. This is presented in Section A1. There are necessary only addition, subtraction, multiplication of complex numbers and the complex plane. If you know these things, you can use the section A1 roll over safely.

A1. The complex numbers

A complex number is in the form:

$$z = a + b * \sqrt{-1}$$

where a and b are common (real) numbers. Since you can not take the square root of a negative number, we take for $\sqrt{-1}$ the value i , ie:

$$z = a + b * i$$

Apparently: $i * i = i^2 = -1$.

Examples of complex numbers:

w = 3 + 5i;   v = -2.4 + 7.8 i;   z = 1-8 i

Complex numbers can be added, subtracted, multiplied and divided. For the production of the Mandelbrot set, we only need the addition and multiplication:

Addition: If $z = a + b * i$ and $w = u + i v *$,

Then it is

$$z + w = (a + u) + (b + v) * i$$

Example: $(2 + 3i) + (1-5i) = 3-2i$

Multiplication: If $z = a + b * i$ and $w = u + i v *$, then

$$z * w = a*u + a*v*i + b*i*u + b*v*i^2 \text{ (multiplying)}$$

Summarize and consider that $i^2 = -1$ yields,

$$z * w = (a*u - b*v) + (a*v + b*u) * i$$

Example: $(3i + 2) * (1 + 2i) = -4 + 7i$

A2. The construction of the Mandelbrot set

All complex numbers you can draw into a plane and this is the Gaussian number plane.

It is a common coordinate system with x-axis (abscissa) and y- axis (ordinate). A complex number $z = x + y * i$ is entered in this plane at the point (x, y). Apparently, any complex number is located in a point of the plane and each point of

Where does the space end?

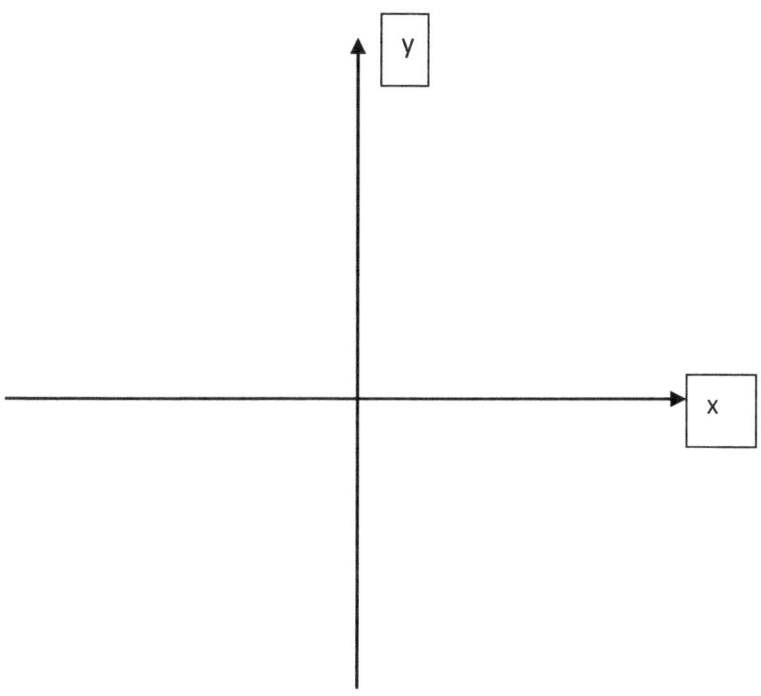

Complex plane

the x-y-plane represents a complex number.
The term

$$|z| = \sqrt{x^2 + y^2} \quad \text{(Pythagoras)}$$

is evidently the distance of a complex number (ie its point in the plane) from the origin of the coordinate system.

$|z|$ is called the amount of the complex number z.

We can now construct the Mandelbrot set in the complex

plane. We simply need to assign to each point (pixel) in the plane a colour. When all points are "coloured", the image (Mandelbrot set) is created.

The assignment of a colour to a point (pixel) with coordinates x and y, ie to the point $z = x + y * i$, is done as follows:

1. Choose $w_0 = 0$

2. Calculate $w_1 = w_0^2 + z$, then $w_2 = w_1^2 + z$, then $w_3 = w_2^2 + z$, etc,

   So $w_{k+1} = w_K^2 + z$ for k = 1,2,3,4 ....

3. We get the number sequence w1, w2, w3, etc.

4. Consider the sequence of the amounts $|w_1|$, $|w_2|$ | w3 | , | W4 | , | W5 | , , .

We consider a high value, for example the hundredth $|w_{100}|$ or the five hundredth $|w_{500}|$, Depending on the size of this value we choose a colour for the set point $z = x + y * i$, so for the pixel (x, y).

Here is an example when w_{100} is the hundredth value:

$|w_{100}|$ <50 ➔ colour red for z
(ie for the point (x, y) at $z = x + y * i$)

50 ≤ $|w_{100}|$ <100 ➔ colour green for z

100 ≤ $|w_{100}|$ <500 ➔ colour blue for z

500 ≤ $|w_{100}|$ ➔ colour yellow for z

When you perform the above procedure in the computer for all points z of the screen area (the Gaussian plane), we obtain the Mandelbrot set.

Particularly aesthetic images are delivered by small surface portions at the edges of the Mandelbrot set (see. Fig.4). If you select very small areas on the edge and increase it, taking into consideration only the complex numbers in this area and extend to the whole Gaussian level, these images (Figure 5, see.) arise

Literature

Brandon Carter
The Constants of Nature
London, 1983.

Albert Einstein and Hanoch Gutfreund
Relativity: The Special and the General Theory
100th Anniversary Edititon, 2015

Werner Kinnebrock
Bedeutende Theorien des 20.ten Jahrhunderts
München, 4.Auflage, 2013

Werner Kinnebrock
Mikro und Makro. Von Galaxien und Aromen
Eine physikalische Reise
München, 2014

Norman K. Glendenning
After the Beginning
A Cosmic Journey through Space and Time
Imperial College Print, 2004z

Andrew Liddle
An Introduction to modern Cosmology
Wiley, 2007

Roger Penrose and Stephen Hawking
The Nature of Space and Time
Princeton Science Library, 2015